99 WAYS TO USE YOUR OSCILLOSCOPE

By Albert C.W. Saunders

TAB BOOKS Inc.
BLUE RIDGE SUMMIT, PA. 17214

FIRST EDITION

SIXTH PRINTING

Printed in the United States of America

International Standard Book No. 0-8306-7483-7

Library of Congress Card Number: 68-56094

Preface

The oscilloscope is an extremely versatile instrument, limited in what it will do only by the user's understanding of its operation. While I have not devoted any of the content to basic fundamentals of scope operation (my previous book, "Working with the Oscilloscope," contains this information), you will find within these pages many worthwhile and unique uses for the oscilloscope—including circuit troubleshooting, receiver alignment, and test procedures—particularly in the realm of new circuits, innovations, and semiconductor devices. Many applications deal with TV circuits, especially color. Actually, the information presented here should help the serious technician or user become increasingly efficient in oscilloscope operation, suggesting numerous additional methods of adapting the instrument's versatility to his everyday tasks.

More and more, as the state of the art develops, the oscilloscope is becoming a necessary and mandatory piece of servicing gear. It gives the user eyes to "see" what's going on in complex circuitry, helping him to reduce even the most intricate devices to their basic circuits and facilitate the diagnosis of component malfunctions. The oscilloscope, in conjunction with a signal generator or signal source, provides an ideal means of performing overall circuit checks. By applying the test signal to the input and connecting the oscilloscope to the output you can immediately assess the condition of any piece of equipment. Then, by using a simple process of elimination, you can trace the trouble to a specific section, circuit, and component. The waveform contour—its amplitude, frequency, and phase—presents a wealth of information to the technician, if he knows what he *should* see! And that's the entire purpose of this book—to help you learn how to use the oscilloscope and how to interpret what you see displayed on the screen.

A. C. W. Saunders

CONTENTS

The Vertical INPUT post:
This is the post ordinarily used for observing wave forms in electronic equipment.

Horizontal Gain:
The horizontal Gain control adjust the amount of signal applied to the Horizontal.

Introduction

The capabilities and limitations of an oscilloscope generally depend on the vertical amplifier characteristics. If usable up to 500 kHz it is considered "narrow band" and up to 4 MHz or higher it is wideband. These ratings are acceptable regardless of sweep oscillator range. Thyratron oscillators usually are limited to 30 kHz; however, conventional sawtooth multivibrators are very popular and provide higher linear sweep rates. For example, a sweep rate of 400 kHz will display a single cycle of a 400 kHz signal. Higher sweeps are obtainable, and some oscilloscopes have a 5-to-1 sweep expansion circuit (see Diagram A); therefore, if a 5-cycle display is the limit for a particular sweep, switching to expansion will produce a single cycle. This is a welcome addition in some critical tests.

Another good feature is the convenience of selecting AC or DC vertical and/or horizontal amplifiers. This feature alone offers many advantages, including the elimination of a phase shift existing between the two amplifiers and improved vertical amplifier low-frequency response. The waveforms shown in Diagrams B and C are actual photographs of 3.58-MHz signals displayed on a wideband oscilloscope. Oscillogram B shows two damped waved trains, the first representing a low Q circuit and the second a high Q circuit. This display was produced by shock existing a bandpass amplifier (color TV) with a flyback pulse. It displays the effect of ringing while the tube was cut off (high Q) and conducting (low Q). To obtain distinct peaks it was necessary to carefully adjust the intensity and focus controls. Stability of the pattern was excellent for time exposure with a regular 35mm camera. Oscillogram C shows the two 50-volt 3.58-MHz color subcarriers in quadrature for reinsertion at the color demodulators. For this display a dual trace oscilloscope was used (with a built-in electronic switch).

Probes

A probe cable is used as a means of coupling the signal to be observed to the vertical input of the oscilloscope. It must be designed for good transfer characteristics. A description of the three most commonly used probes follow:

DIRECT PROBE: Most direct probes contain a medium shunt capacitance and are suitable for testing low-impedance cir-

A

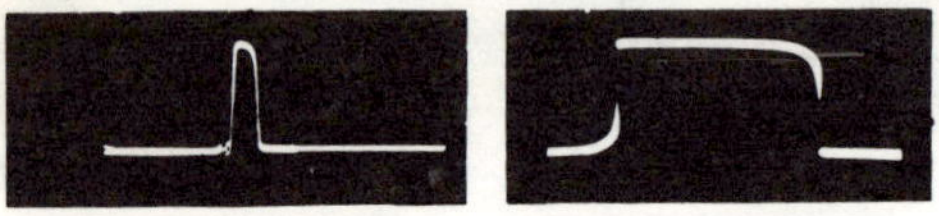

5 to 1 horizontal expansion

B

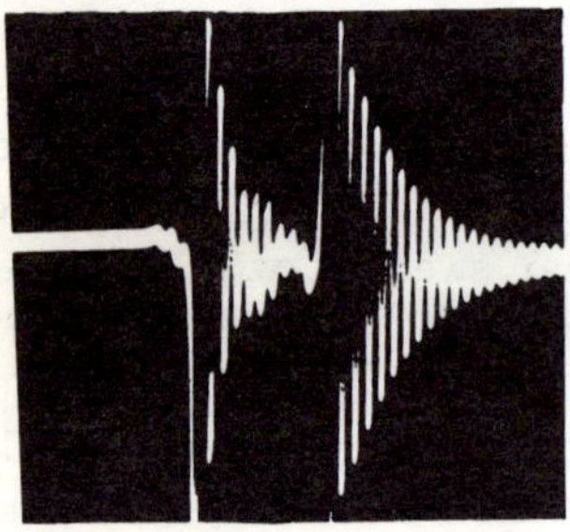

Damped waves from shock excited 3.58 MHz amplifier.

C

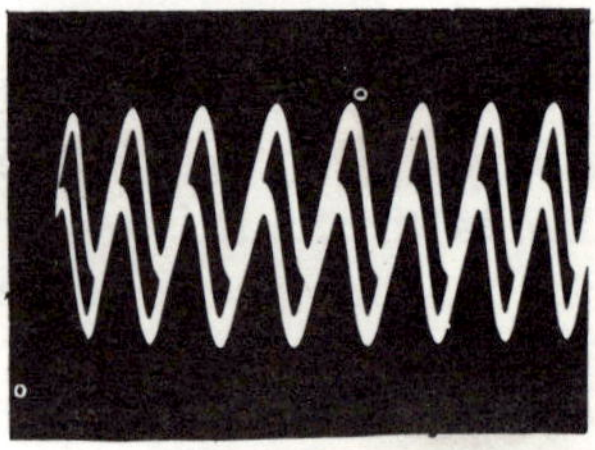

Two - 3.58 MHz color subcarriers in quadrature. Electronic switch.

cuits such as audio waveforms in both vacuum tube and transistor equipment. A good rule to follow when testing equipment is: "The impedance of the testing device must be at least ten times greater than the circuit it shunts." The greater the better, but a 10-to-1 ratio does not affect the constants of the circuit under test.

LOW-CAPACITY PROBE: A low-capacity probe is used for observing waveforms across high-impedance circuits and it is practically a must in must TV circuits. The probe consists of a megohm series resistor shunted by a 5 pfd to 25 pfd variable capacitor and terminated by a 100K resistor. This gives the probe an attenuation ratio of 10 to 1, which is designed to operate in conjunction with the oscilloscope decade step attenuator. The probe is permanently calibrated by applying a 10-kHz square wave and adjusting the small capacitor in the probe until a good square wave is displayed. A probe of this type should NOT be used on potentials higher than 500 volts.

DEMODULATION PROBE: Sometimes referred to as a wandering detector, a demodulation probe is used chiefly in checking RF, IF, and FM tuned circuits to observe a response curve from a sweep generator. The response of a single stage, or two or more in cascade, may be checked with a

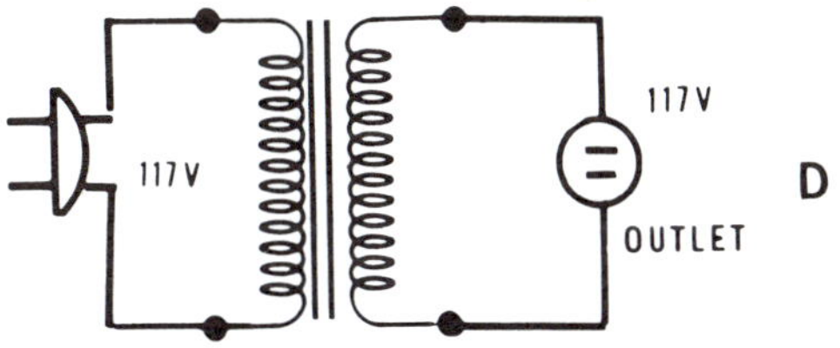

ISOLATION TRANSFORMER

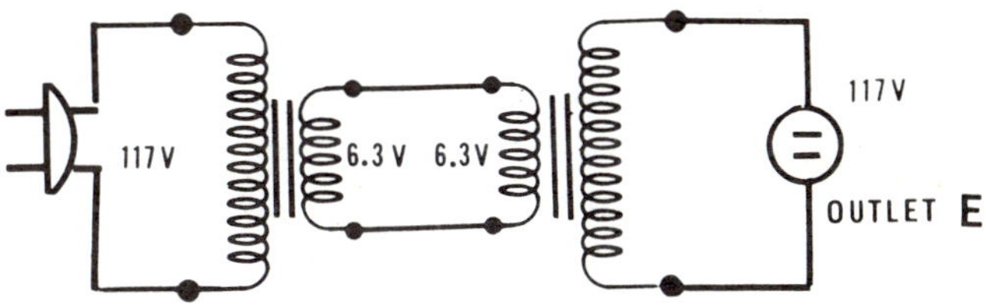

ISOLATION TRANSFORMER USING
TWO FILAMENT TRANSFORMERS..

demodulator probe. It may also be used on RF and IF amplitude-modulated signals to observe the audio or video component in signal tracing. (For viewing the waveforms at the output of the horizontal output and the input of the HV rectifier the author has suggested two proximity tests in the TV section.)

Safeguarding Personnel and Equipment

Certain precautions must be observed when setting up test equipment with AC/DC operated units. Usually, one side of the power line is grounded in such equipment; therefore, it is a good practice in any type of test to use an isolation transformer with a wattage rating adequate to handle the load. See Diagram D. By operating without such protection in any test there is a possibility of the grounded side of the line making contact with the HOT side by cross connection of test leads or chassis touching chassis, thus causing a short circuit or possible damage.

As an expedient, two filament transformers connected as shown in Diagram E serve as an ideal isolation transformer, since it offers double isolation. However, the wattage rating of both transformers must be capable of handling the load. Just connect it as shown in the diagram; no phasing is necessary. The ground terminals of an oscilloscope are common to its housing; therefore, care must be exercised when checking waveforms where ground is not a reference point. Metal benches should have a good masonite cover with or without isolation transformers.

Pertinent Points

The ground terminals of both vertical and horizontal inputs are common to one another and either one may be used for ground purposes. The terminals also are common to the oscilloscope housing, and care should be taken when checking voltages and waveforms where ground is not used as a reference point; for example, checking across any one of the deflection coils in a TV receiver. Under these test conditions the oscilloscope housing is charged to approximately 300 volts. It is advisable to use a DC blocking capacitor in the ground lead.

When using the high-sensitivity ranges it is a good practice to use a well shielded probe and cable. A shield cover over the terminals is sometimes necessary in noisy locations. Connector cables that automatically ground the shield lead are considered ideal for high-impedance tests. The manufacturer specifies the maximum allowable potential that be applied to the vertical amplifier input. This precaution should be observed at all times! Potentials higher than those specified may be checked by using an attenuator probe, available in 10-to-1 and 100-to-1 ratios, designed to operate in conjunction with the vertical amplifier decade attenuator.

Care must be exercised when checking a TV flyback transformer. Direct contact with conventional instruments should NOT be made at the plate of the horizontal output tube or the high-voltage rectifier. Capacity voltage dividers are practical but not always convenient. In preparing for tests in the HV section always switch off the power.

All RF signal generators, whether CW, modulated, or sweep, should be terminated with an appropriate carbon resistor. For the correct value consult the manufacturer's operating manual.

1

Calibration and the Graticule

Oscilloscopes designed for general service work are usually provided with a built-in voltage source for calibration purposes, such as a "one-volt" peak-to-peak test signal. The more expensive instruments have a calibrated step attenuator for both vertical and horizontal amplifiers and a graticule that divides both the vertical and horizontal axes into centimeters. See Fig. 1. Notice that the vertical sensitivity control is calibrated in steps from .01 volt/cm to 10 volt/cm. The horizontal sweep attenuator is calibrated in 15 steps from 5 μsec/

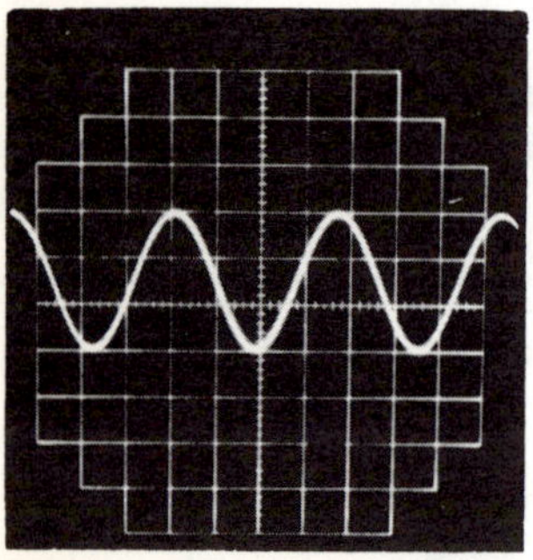

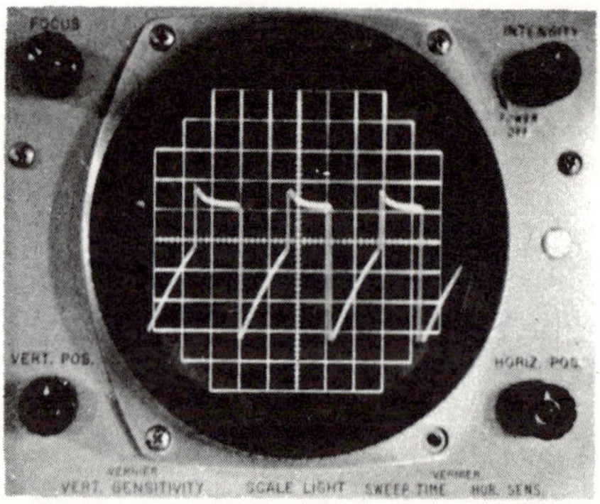

cm to 200 millisec/cm. The same control also provides for external horizontal sweep "horiz sens" which is calibrated in three steps from .1 volt/cm to 10 volt/cm.

Both axes also may be calibrated to measure current, frequency, and angular degrees. A provision also is made for expanding the horizontal trace to measure very small fractions of the practical unit—for example, measuring time in "nano seconds." These calibration procedures are discussed in the following pages. An additional feature, if not already provided, is a low-voltage source terminal on the control panel;

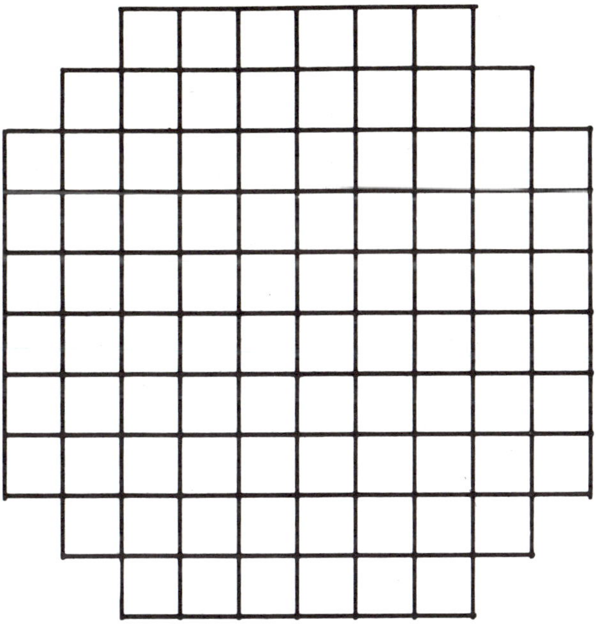

GRATICULE

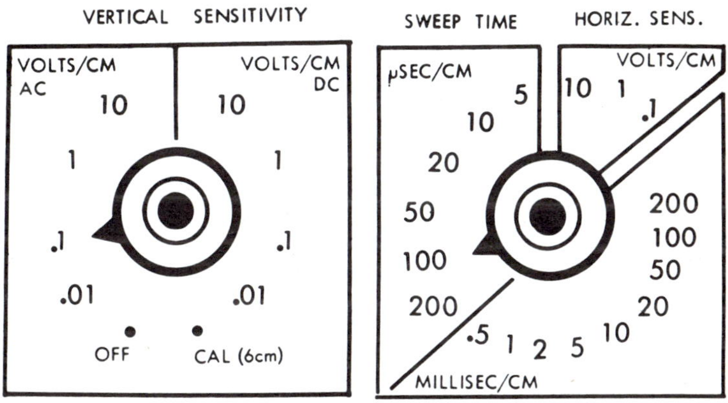

CALIBRATED ATTENUATORS

FIG. 1

besides serving as a test signal it may be used for 60-Hz line sync. A 60-Hz test signal may be tapped from one side of the 6.3-volt filament circuit and connected through a large enough capacitor to the panel terminal.

2

Calibrating the Oscilloscope with an AC Voltmeter

An oscilloscope without a built-in test signal may be calibrated by using a known voltage. The circuit in Fig. 2 will provide such a signal.

Procedure

Equipment Required:

Oscilloscope
Filament transformer (6.3v)
AC voltmeter
3K potentiometer

Shunt the 6.3-volt transformer secondary with the 3K potentiometer and the AC voltmeter as shown in Fig. 2. Switch the sync selector to "line sync." Turn the horizontal gain control to zero and adjust the vertical gain control to produce a vertical trace. Adjust the potentiometer to 6 volts and the vertical gain control to provide a 3-inch trace. This represents 6 volts RMS; therefore, the calibration is 2v/inch RMS. Without changing the vertical gain control setting, proceed to measure the unknown value. If the unknown voltage shows a 2-inch trace this represents 4 volts RMS. To find peak and peak-to-peak values, use the following equations:

$$P = RMS \times 1.414$$

$$P/P = RMS \times 1.414 \times 2$$

Instead of using the straight line vertical trace, turn up the horizontal gain and view a complete cycle. See Diagram A (horizontal sweep rate 60 Hz).

FIG. 2

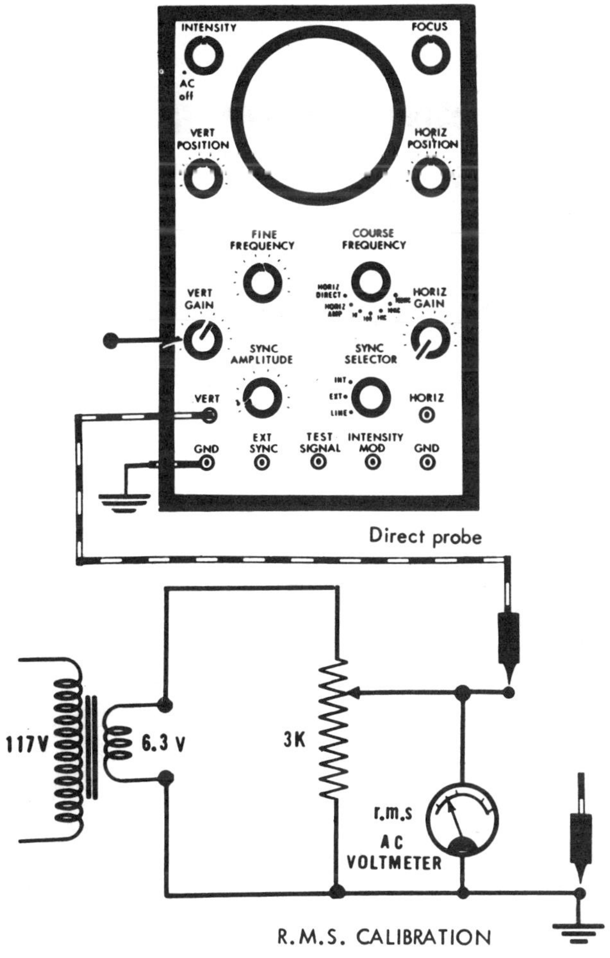

R.M.S. CALIBRATION

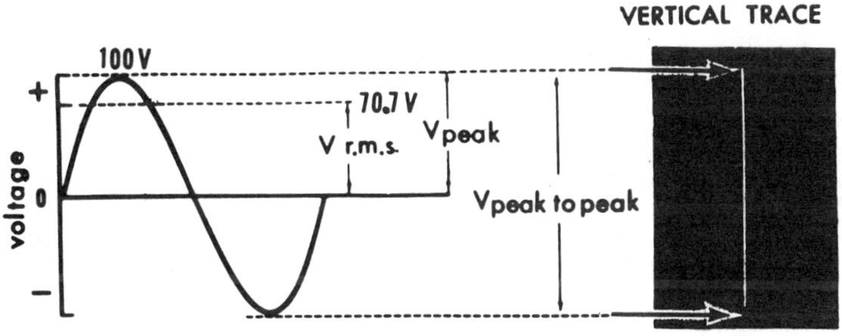

DIAGRAM A

3

Checking Frequency Response Using a Square Wave

The waveform shape observed on the screen is no better than the oscilloscope's ability to reproduce it; for example, the waveform applied to the vertical amplifier may be a near perfect square wave, but due to some malfunction in the oscilloscope or signal generator output connection, the waveform may consist of curves, tilts, or other distorted features. Considering these possibilities a square-wave test should be the first step in any test procedure. It is possible also that the oscilloscope is operating efficiently, but the square-wave generator is at fault.

Procedure

Equipment Required:

Oscilloscope, low-capacity probe
Square-wave generator

Connect the equipment as shown Fig. 3. Tune the square-wave generator to 50 Hz and adjust the oscilloscope sweep rate to display three or four cycles. In this test the waveform may show a slight tilt, top and bottom. This is more pronounced when the generator is tuned to 30 Hz, but at 50 kHz some curvature in the rise and fall on the leading and trailing edges will appear. At 500 kHz the curvature will be very pronounced (see the oscillograms in Fig. 3). A good square wave contains the 100th harmonic of its fundamental. For example, if a good square wave is produced at 1 kHz then the response of the oscilloscope is flat to a sine wave response of 10,000 cycles (10 kHz). The repetition rate of the square wave is the first harmonic. Rise time of the leading edge is measured from 10% to 90%. On wideband oscilloscopes fairly good square waveforms may be obtained up to the MHz range.

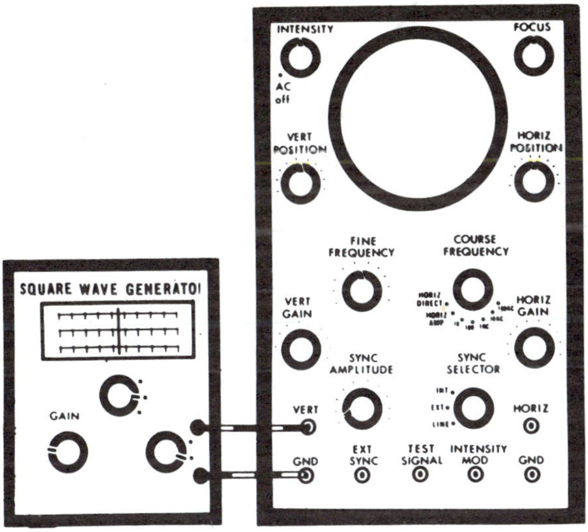

5 KHz

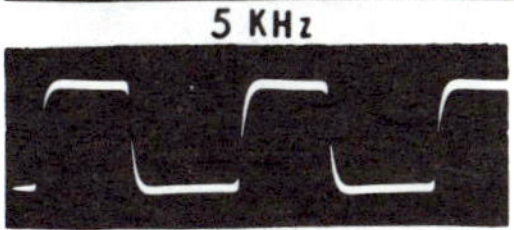

50 KHz

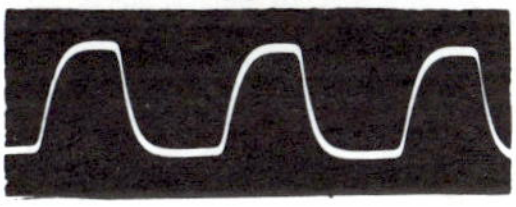

500 KHz

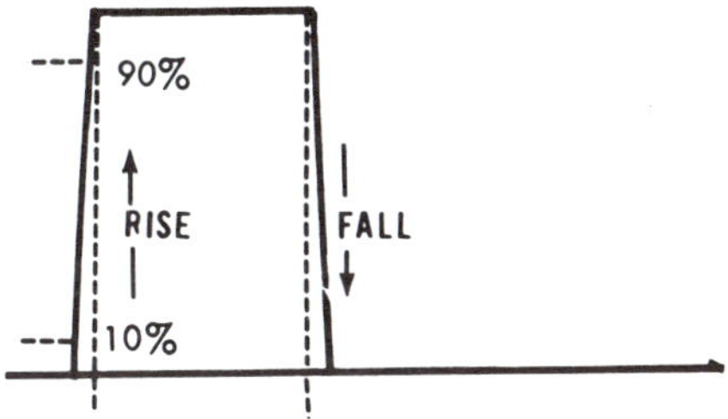

FIG. 3

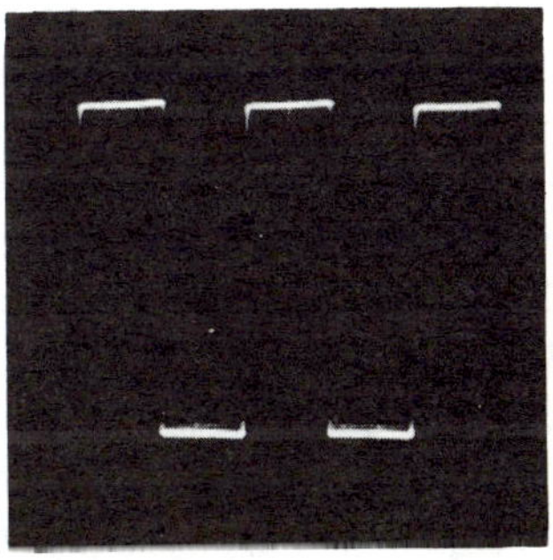

Leading & trailing edges are not visible due to rapid trace

4

Inexpensive Voltage Calibrator

An inexpensive 6-volt peak-to-peak standard for calibrating an oscilloscope is shown in Fig. 4. It consists of a simple transistor amplifier that requires a one-volt peak-to-peak signal from an audio signal generator or equivalent.

Procedure

Equipment Required:

Oscilloscope
Audio generator
Test circuit

Connect the equipment as shown in Fig. 4 and tune the audio signal generator to about 100 Hz. Adjust the oscilloscope sweep rate to display a few cycles. Turn up the generator volume control until pulses appear at the collector; advance the volume control further until the pulses reach the maximum peak. It will require about a 1-volt peak-to-peak sine wave. This may be checked at the generator output. The transistor operates as a switch and converts the battery potential into 6-volt peak-to-peak pulses which are used for calibration purposes. The amplifier is designed to reach saturation and cutoff rapidly. The accompanying oscillogram is a dual trace showing the input sine wave and the output pulses simultaneously.

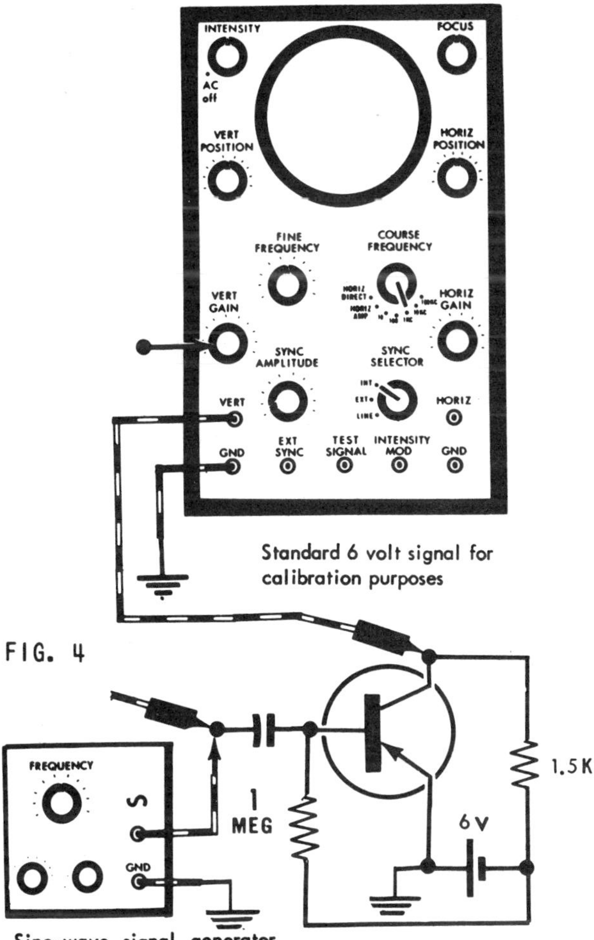

Standard 6 volt signal for calibration purposes

FIG. 4

Sine-wave signal generator

Sine-wave to square-wave

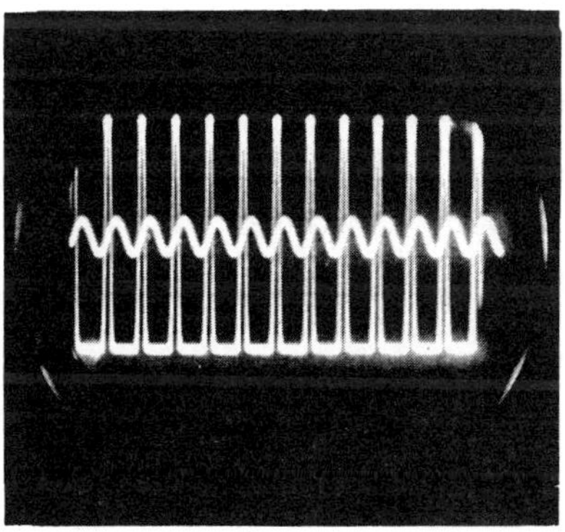

5

Checking Datum Line for Zero Reference

When measuring the peak voltage of sine and nonsinusoidal waveforms it is necessary to establish a zero-voltage reference or datum line. The following suggests a convenient method of periodcially checking the base line during measurements.

Procedure

Equipment Required:

Oscilloscope
Audio signal generator

When a signal is applied to the vertical input the horizontal reference line disappears and the printed line on the graticule is used as a zero reference for peak measurements. However, a periodic check should be made to make sure that the horizontal sweep has not drifted up or down. One way is to remove the input signal, which is not always convenient and it is time consuming. It is NOT good practice to short-circuit the input terminals; therefore, another means must be devised. A convenient method is to install a push-button switch (push-to-open) in series with the vertical test lead. See Fig. 5. Some shops have incorporated this switch on the oscilloscope control panel located near the vertical input terminal. When the switch is depressed it disconnects the signal and the horizontal reference line appears, enabling the technician to check its position. And by pushing the button in rapid succession the signal and its zero voltage reference seem to appear simultaneously. Checking the datum line is very important when DC oscilloscopes are used. In any case a 10-minute warmup period is recommended.

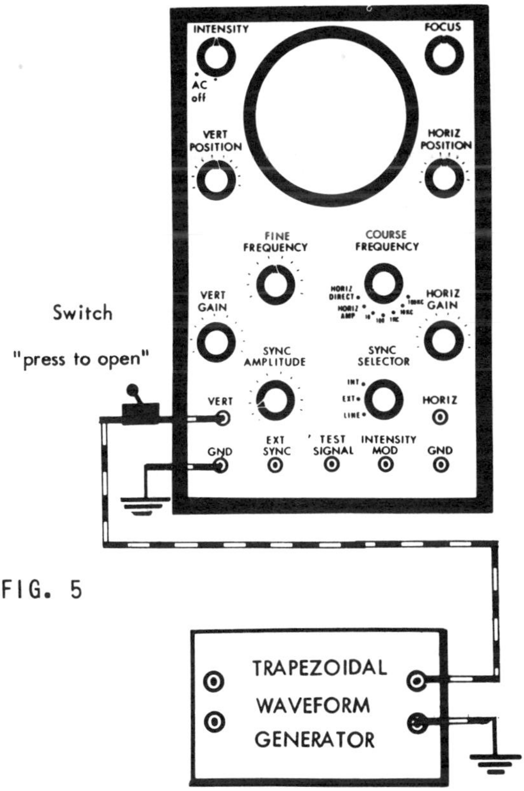

FIG. 5

Measuring the peak to peak voltage of the sawtooth portion of a trapezoidal waveform

Zero reference of a trapezoidal waveform.

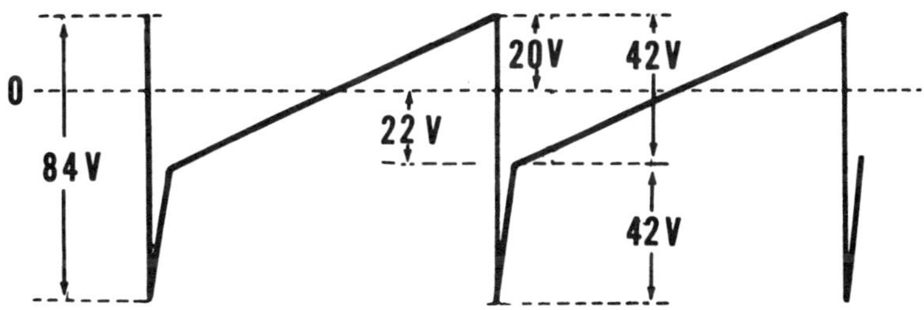

6

Pertinent Waveform Polarity Data

The vertical trace of an oscilloscope conforms with the polarity of a sine wave and a square wave; i.e., POSITIVE-GOING—up trace, and NEGATIVE-GOING—down trace. See Diagram A. The waveform contains information regarding cutoff and saturation of a vacuum tube or transistor stage. A study of the collector waveform of the NPN transistor stage in Fig. 6 shows that as the collector current increases, the collector voltage decreases and is displayed as a "negative-going" or downward trace. Therefore, the negative peaks indicate maximum conduction or saturation. See Oscillogram 2 - Q2.

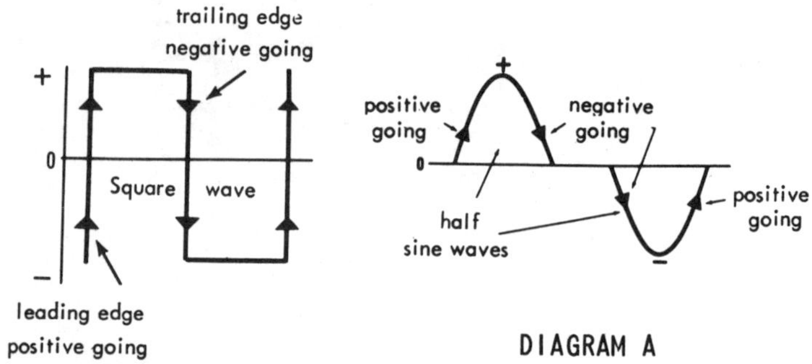

DIAGRAM A

When the collector current decreases, the collector voltage rises and is displayed as a "positive-going" or upward trace, in this case the positive peaks indicate cutoff. Notice the duration of each alternate pulse reveals that the transistor conducts for a short time and is cutoff for a longer period. Using a calibrated graticule it is possible to measure the time periods of cutoff and conduction. The base waveform of Q1 (Oscillogram 1) also conforms with this timing. The discharge curve of capacitor C1 falls from maximum positive to zero during the period Q1 is cutoff since it is a PNP. A further study reveals that Q1 and Q2 cutoff and conduct simultaneously.

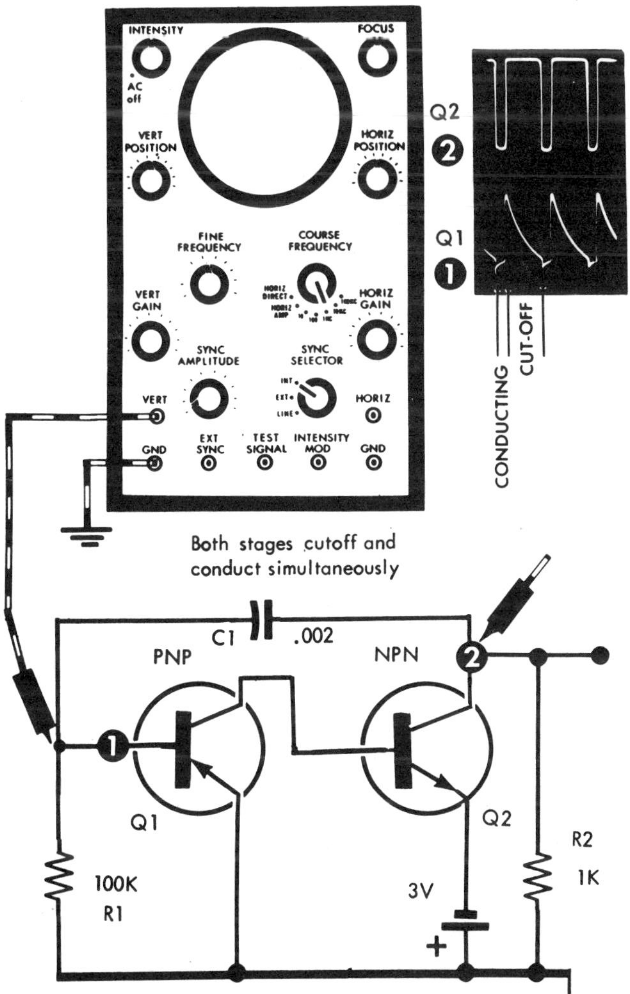
INTENSITY
AC
off
FOCUS
VERT
POSITION
HORIZ
POSITION
FINE
FREQUENCY
COURSE
FREQUENCY
HORIZ
DIRECT
HORIZ
AMP
VERT
GAIN
HORIZ
GAIN
SYNC
AMPLITUDE
SYNC
SELECTOR
INT
EXT
LINE
VERT
HORIZ
GND
EXT
SYNC
TEST
SIGNAL
INTENSITY
MOD
GND
Q2
2
Q1
1
CONDUCTING
CUT-OFF
Both stages cutoff and
conduct simultaneously
C1
.002
PNP
NPN
1
2
Q1
Q2
100K
R1
3V
+
R2
1K

FIG. 6

7

Determining Frequency with Lissajous Figures

Quite often it is necessary to measure an unknown frequency, and Lissajous patterns offer a simple method of determining frequency by comparison with a known standard. A simple introduction to this technique is the use of the 60-Hz line frequency as a standard to check the low-frequency range of an audio generator. Although the reference is fixed at 60 Hz it will serve to introduce the significance of Lissajous patterns.

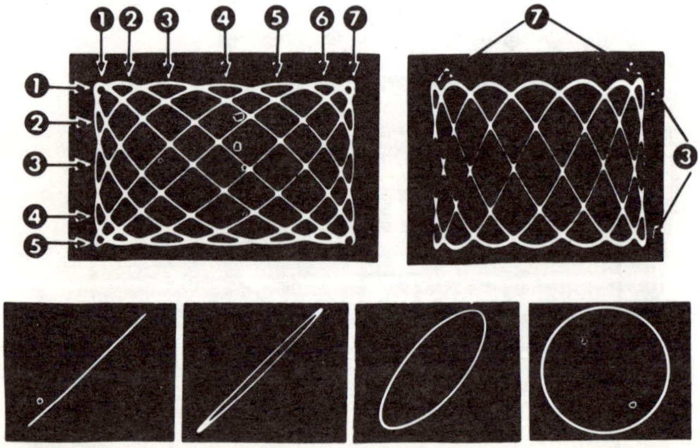

0° 15° 45° 90°

Procedure

Equipment Required:

Oscilloscope
Audio signal generator
Filament transformer (6.3v)

Connect the 60-Hz test signal (or alternate 6.3-volt circuit) to the horizontal input of the oscilloscope and the output of the

FIG. 7

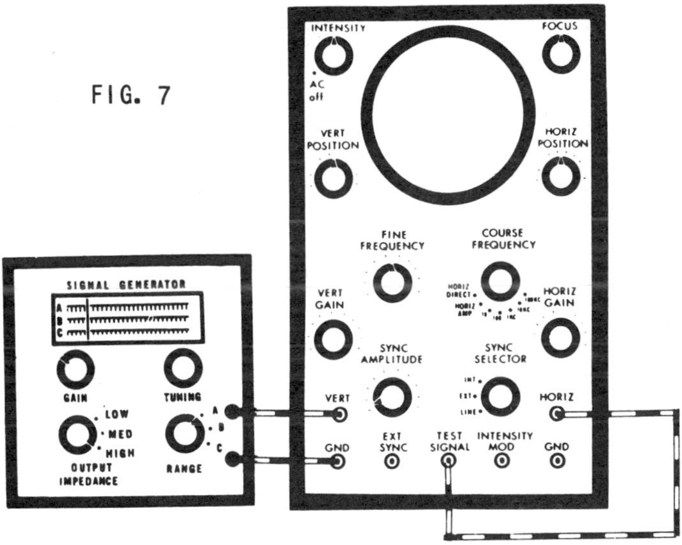

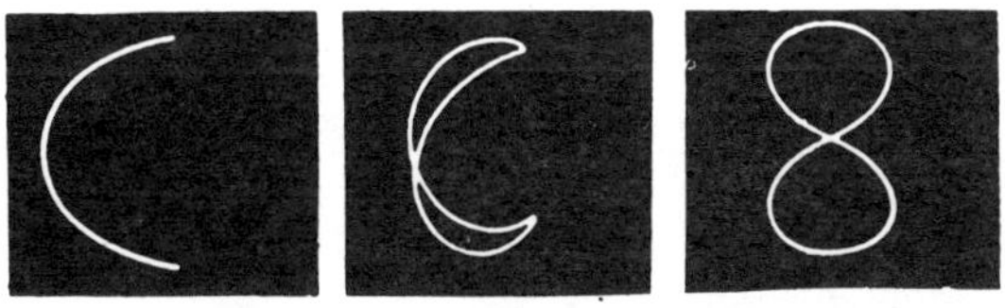

fy/fx = 1/2 (pattern revolving)

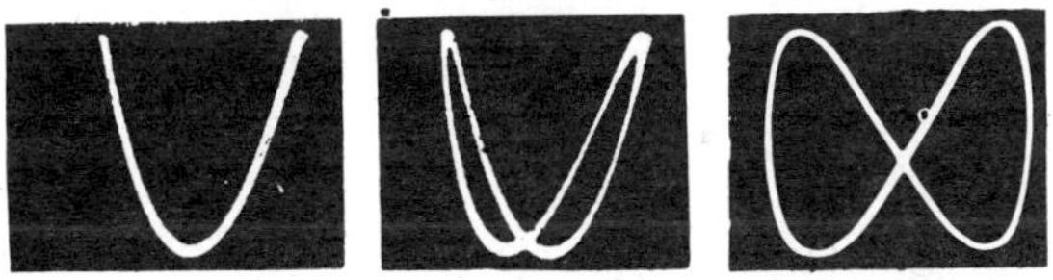

fy/fx = 2/1 (pattern revolving)

fy/fx = 1/3 (pattern revolving)

fy/fx = 3/2 (pattern revolving)

audio oscillator to the vertical input. See Fig. 7. Turn sweep rate selector to "horiz amp" and adjust the gain control to produce about a 3-inch sweep. Turn the audio oscillator to 60 Hz and adjust the vertical gain control to produce a line that slants 45°. The pattern will be revolving on two axes due to phase shift. This is to be expected since no sync voltages are involved. Hence the pattern will shift from a slanting line to an ellipse and then to a circle. When this condition exists the vertical frequency (fy) is equal to the horizontal frequency (fx), or fy = fx. Tune the oscillator to 90 Hz, producing a ratio (fy/fx) or 90/60 or 3 to 2. The oscillograms show several ratios.

8

Calibrating the Time Base

To measure time, some provision must be made to calibrate the horizontal sweep. Here is a simple, but effective, method.

Procedure

Equipment Required:

Oscilloscope
Audio signal generator

Connect the signal generator output to the oscilloscope vertical input terminals. Tune the signal generator to 1 kHz and adjust sweep frequency and horizontal gain to display one cycle as illustrated in Fig. 8. Notice that the cycle engages 10 horizontal divisions, which represents 1000 microseconds or 100 microseconds per division. The chart indicates the time period per cycle for several frequencies. If absolute accuracy is necessary, the frequency used for calibration must be crystal controlled. A slight plus or minus frequency shift due to heat can be tolerated. A very good low-frequency standard is the 60-Hz line frequency, where one cycle engaging 10 divisions represents a time base of 1/60 of a second. A 6.3-volt filament transformer, connected to the oscilloscope vertical input, offers a calibration voltage source.

FIG. 8

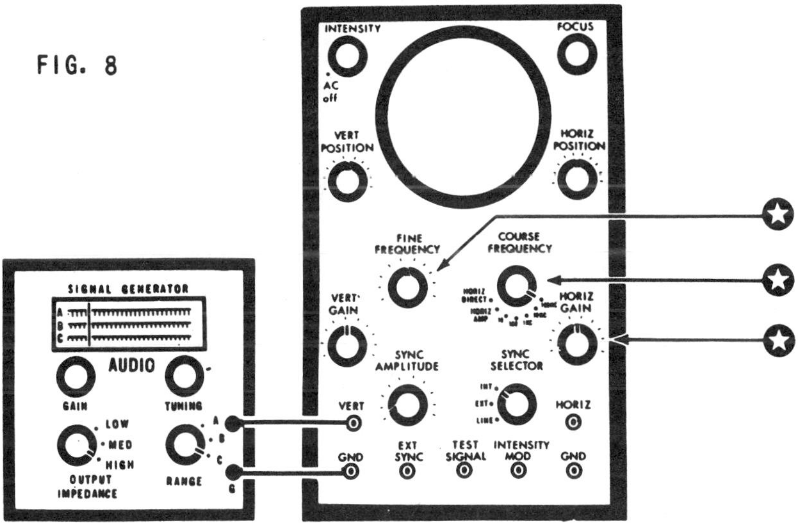

TUNED TO 10 KC

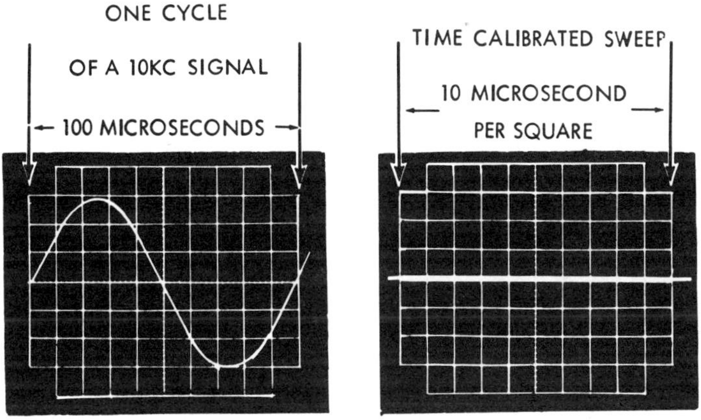

SIGNAL	TIME PER CYCLE
1 KHZ	1000 microseconds.
10 KHz	100 microseconds.
100 KHz	10 microseconds.
1 MHz	1 microsecond.

9

Measuring Inductance (L) and Inductive Reactance (XL)

Let us assume that both L and X_L of a particular coil are unknown. We must, of course, determine X_L at some particular frequency, and this can be accomplished by using a relatively low AC voltage source such as a 1-kHz signal from an audio generator. Fig. 9 shows the use of a 60-Hz low-voltage source, as an example.

Procedure

Equipment Required:

Oscilloscope
DPDT switch
Audio signal generator

Connect test circuit as shown in Fig. 9. Switch to position X_L. Adjust the oscilloscope display to two cycles with a vertical deflection of four divisions. Switch to position R and adjust the series resistor to obtain the same vertical deflection. Switch back and forth from X_L to R and readjust until both patterns are equal in height. Measure the active portion of R with an ohmmeter. Since R and X_L have the same AC voltage drop, the ohms measured across R will be the same for X_L. Since X_L is equal to $2\pi fL$ we can transpose the equation to find the value of L. Assume that X_L is 1000 ohms and the frequency is 1 kHz. Then:

$$L = \frac{X_L}{2\pi f} = \frac{1000}{6280} = .16 \text{ henry (Approx.)}$$

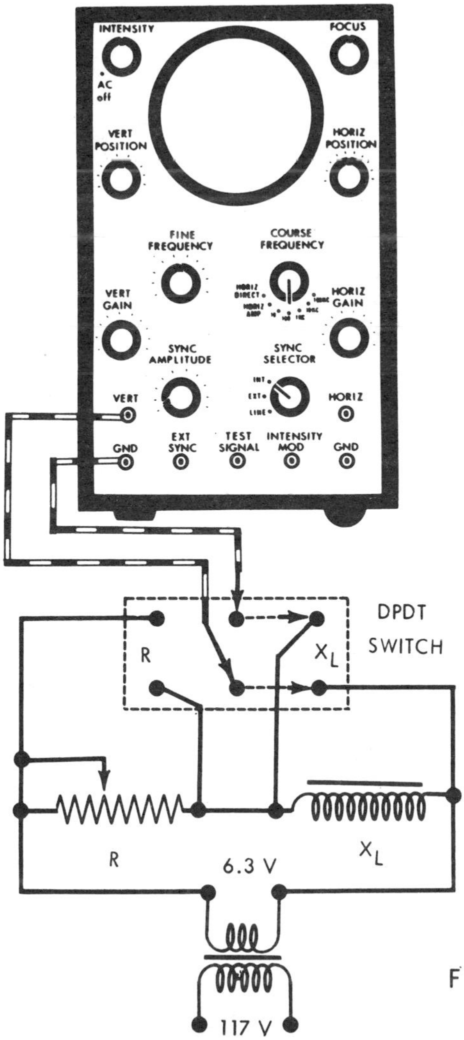

FIG. 9

E_{XL}

E_R

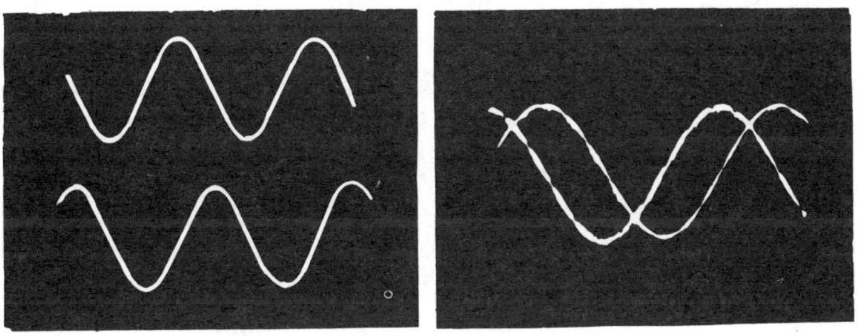

$R = 2\pi FL$ Oscillograms using dual trace

Note 90 degree phase difference between E_{XL} and E_R

10

Zener Diode Curve Tracer

Increasing the reverse bias potential beyond the specified rating causes a Zener diode to breakdown and conduct heavily; a further increase beyond this point causes the diode current to increase while the voltage drop across the diode remains constant at the specified potential. To facilitate the test a 20-volt 1-watt Zener diode was selected and connected as shown in Fig. 10. A 6-watt 117-volt lamp is used for protection against accidental short circuit and a 10K resistor for current limiting.

Procedure

Equipment Required:

Oscilloscope (adjusted to 10 volts per division horizontally)
20-volt 1-watt Zener diode
Curve tracing adapter

Set up the test as shown in Fig. 10. Do not apply AC power until later. Rotate vertical and horizontal gain controls fully counterclockwise to produce a spot and adjust the positioning controls so that the spot is at dead center on the graticule. This position represents zero volts on both axes. Apply AC to the tracer circuit and turn up both gain controls so that the display is similar to the oscillogram. Notice that at right of center the horizontal sweep traverses two divisions and traces vertically. This indicates a breakdown of the reverse bias when it reaches 20 volts (2 divisions). At this point the diode conducts heavily, controlled only by the 10K resistor. Left of center represents the forward bias direction; notice here that in this direction the diode conducts during the positive half-wave swing of the applied cycle indicated by the vertical deflection.

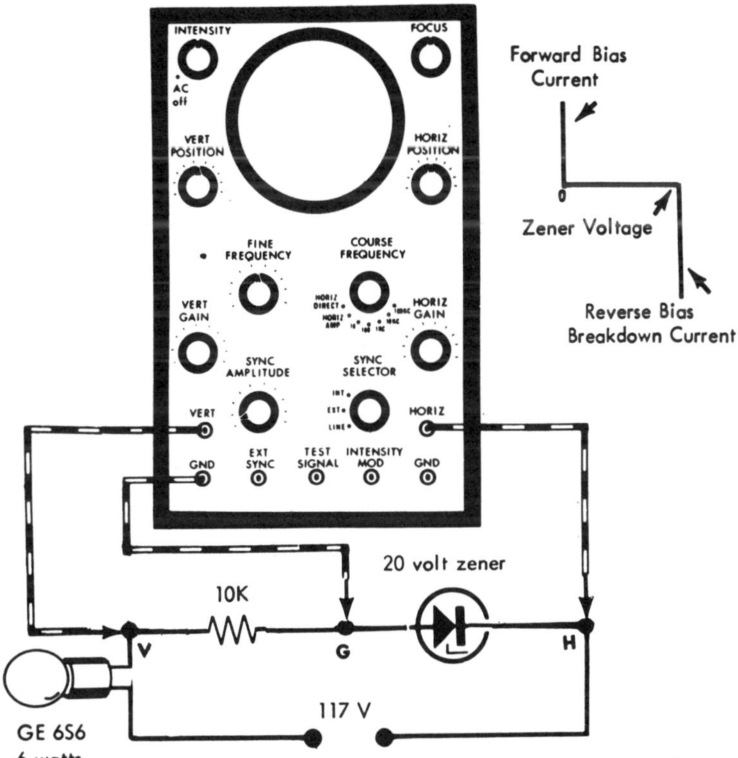

FIG. 10

HORIZONTAL CALIBRATION
10 volt/cm

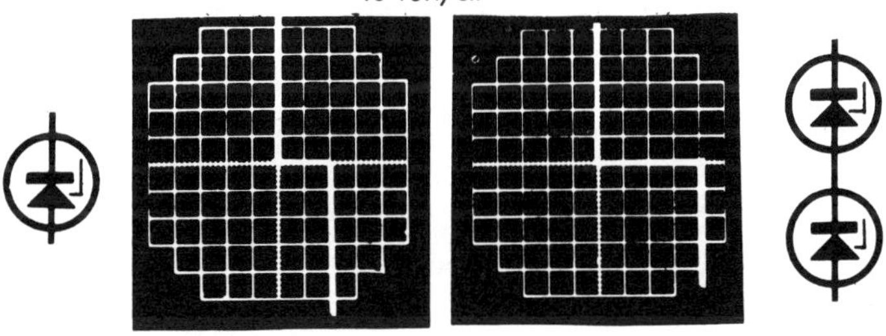

One 20 Volt Zener Diode

Two 20 volt Zener Diodes in series

11

Tunnel Diode Curve Tracer

A forward or reverse bias of less than 50 millivolts will cause a tunnel diode to conduct in either direction. This is, of course, contrary to the characteristics of a conventional diode; however, in the forward-bias direction the diode current increases and then decreases despite an increase of forward bias. This negative resistance characteristic is the important feature of the tunnel diode. However, a further increase of forward bias above 300 millivolts (for germanium) will cause the diode current to increase (see Diagram A). The dotted line represents a conventional diode.

Despite the bi-directional characteristic, the diode does exhibit a negative resistance in the forward-bias direction. Testing a conventional diode with an ohmmeter to determine its forward and reverse resistance ratio is both simple and expedient, since a 10-to-1 ratio indicates a good diode. However, the ohmmeter test MUST NOT BE USED ON TUNNEL DIODES. First, the diode conducts equally in both directions and a resistance test serves no useful purpose. Second, damage to the diode can result, especially on a low resistance (high current) range of the ohmmeter. It is the dip in the characteristic curve that determines whether the diode is normal or defective.

Procedure

Equipment Required:

Oscilloscope
Curve tracer circuit

Connect the oscilloscope to the curve tracer as illustrated in Fig. 11 and adjust potentiometers R1 and R2 to produce a balanced curve as shown in the oscillogram.

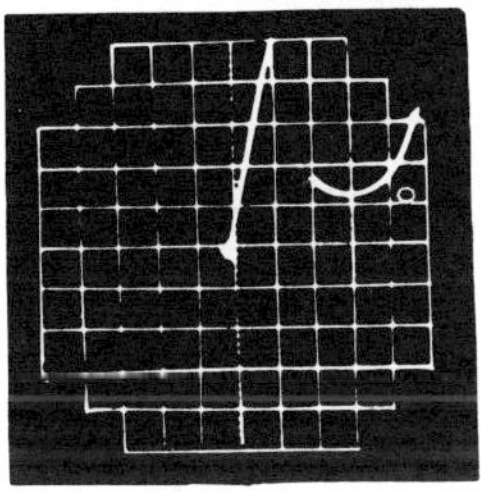

With diode D1

DIAGRAM A

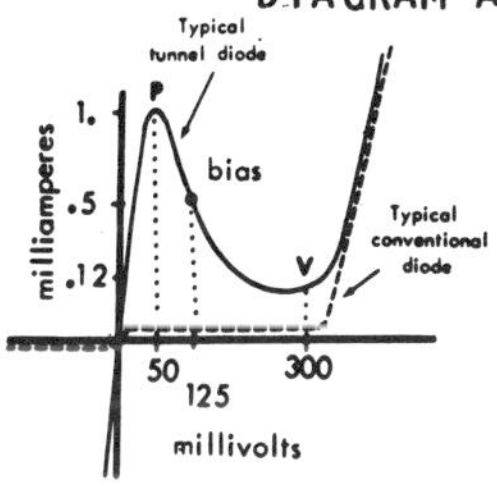

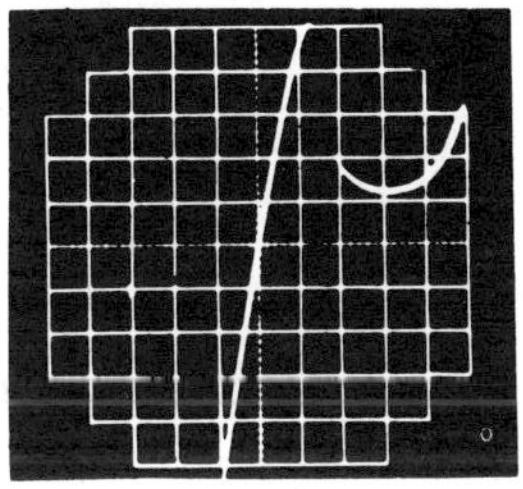

Without

P.....Indicates Peak Point Current & Voltage

V.....Indicates Valley Point Current & Voltage

The slope between P & V represents negative resistance

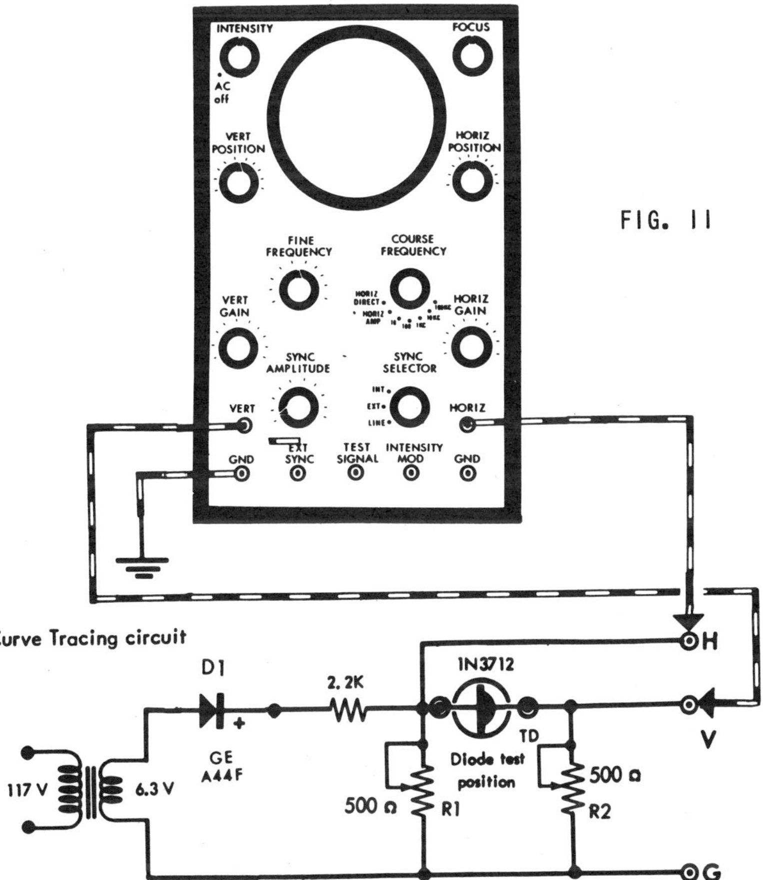

FIG. 11

Controls R1 and R2 are adjusted to produce a balanced curve as shown in the oscillogram.
If a double trace is observed switch vertical and horizontal amplifiers to d.c. operation.

12

PNP Transistor Curve Tracer

In conjunction with an inexpensive test circuit the oscilloscope makes an excellent curve tracer for transistors. See Fig. 12. This arrangement permits the technician to see the transistor in action while varying the base current.

Procedure

Equipment Required:

Oscilloscope
Test circuit

To obtain the characteristic curve, the collector voltage is made to sweep horizontally from zero to -8.5 volts peak. This is accomplished by rectifying the 6.3-volt (RMS) secondary voltage. Hence, the horizontal sweep represents the applied collector voltage which is automatically calibrated from zero to -8.5 volts at the end of the trace. The vertical trace represents the collector current flowing through the 100-ohm resistor in the collector circuit. This trace may be calibrated in volts per vertical division, making it possible to determine the collector current by Ohm's Law.

If by chance an NPN transistor is tested in this circuit it will display a horizontal line. Reverse the collector and emitter leads and it will display a curve similar to a normal diode. No damage will occur to the NPN unit, thus making the tester capable of indicating the difference between PNP and NPN. It is possible to reverse the leads (collector and emitter) when testing a PNP. If this should happen the curve will show very little gain and will remain like this regardless of the amount of base current applied. Remember this when replacing a transistor in a receiver and double check the lead locations.

CURVE TRACER

PNP

TRANSISTOR

FIG. 12

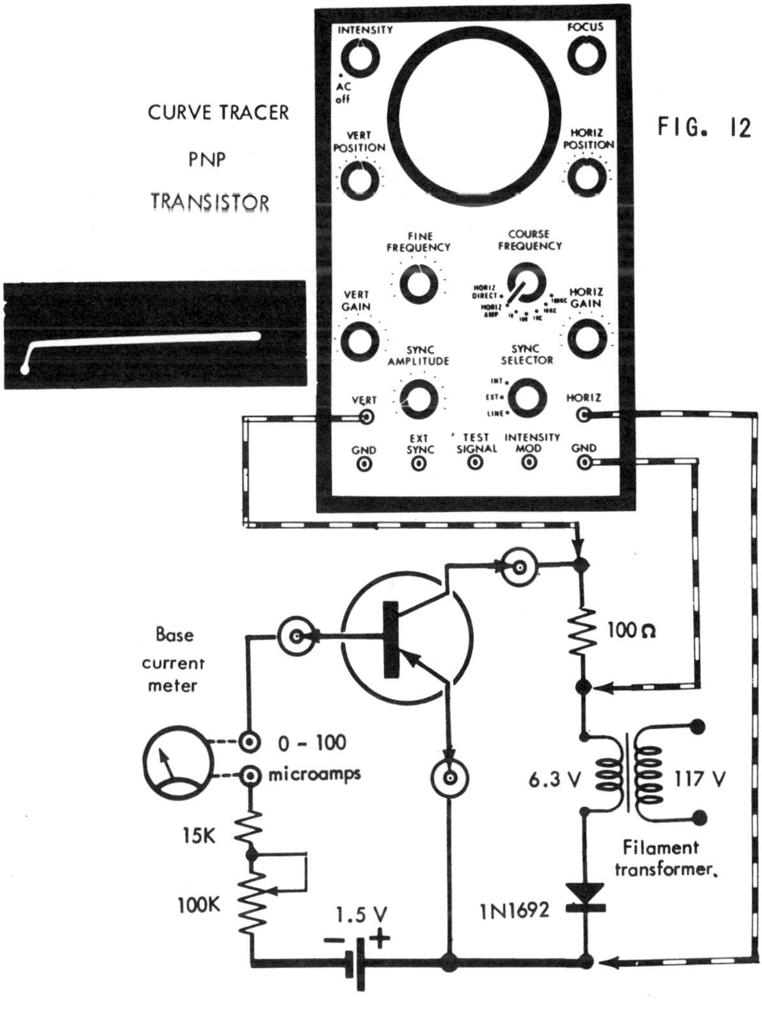

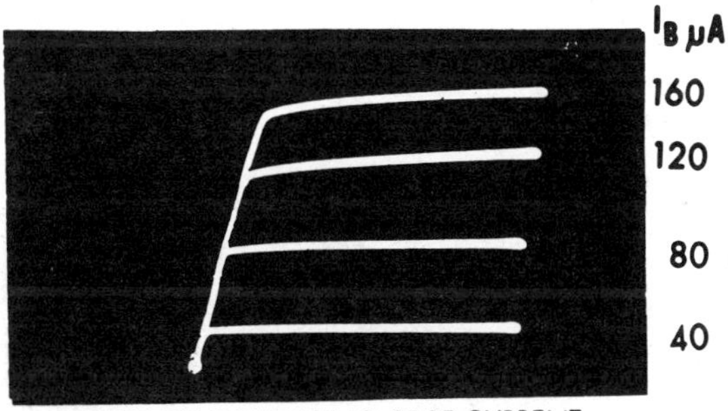

EFFECT OF INCREASING BASE CURRENT

13 NPN Transistor Curve Tracer

The curve tracer in Fig. 13 is similar to that used for PNP types, with the exception that the test polarities are reversed.

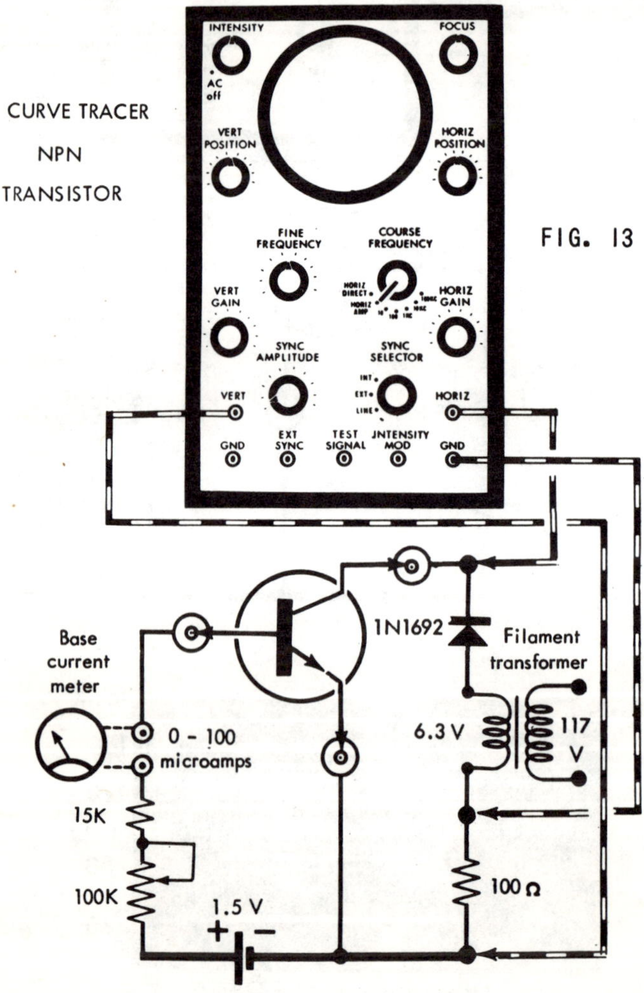

FIG. 13

14

PNP Power Transistor Curve Tracer

The curve tracer shown in Fig. 14 is modified to accommodate a power transistor. Notice the 100-ohm collector resistor has been replaced by a 0.5-ohm (1/2-ohm) unit. Also, the 1.5-volt base bias battery has been replaced with a 6-volt battery.

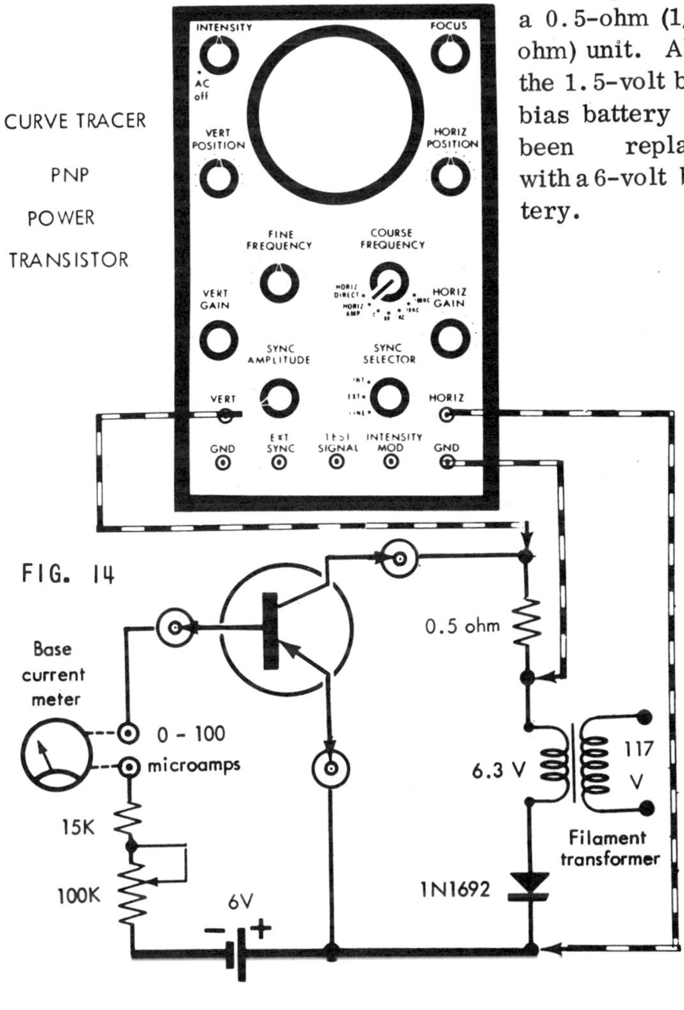

FIG. 14

15
NPN Power Transistor Curve Tracer

Notice that these transistor curve tracer circuits are individual and not combined. This separate unit tester is simple to hook up and does not need double-pole switches. It makes each test simple and expedient.

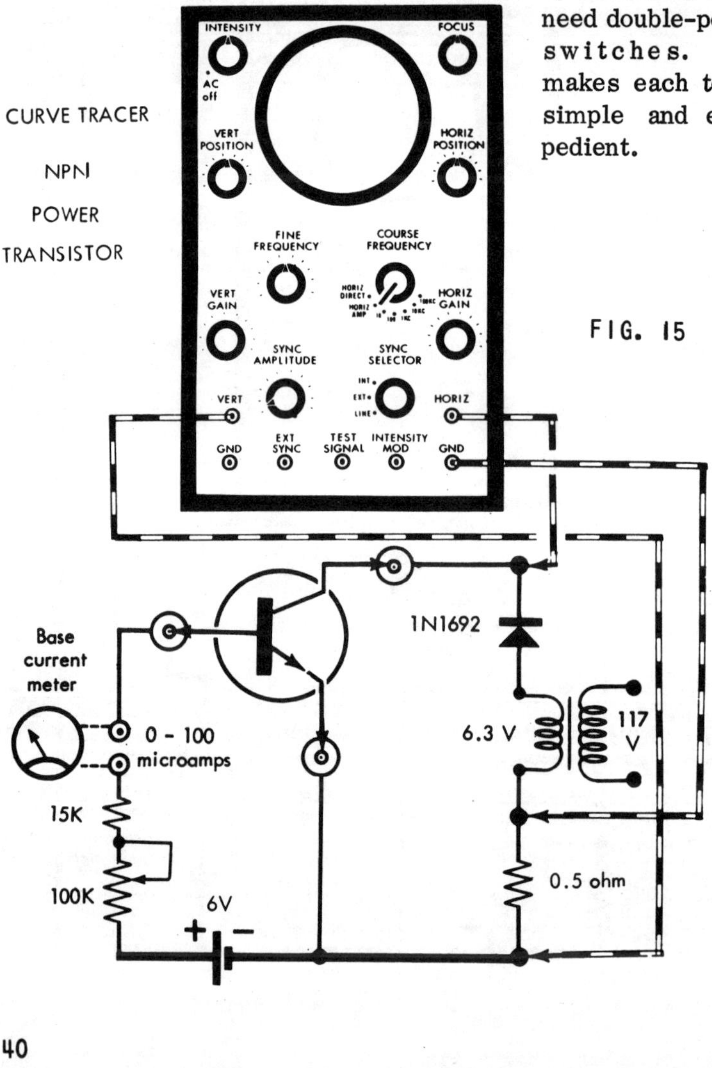

FIG. 15

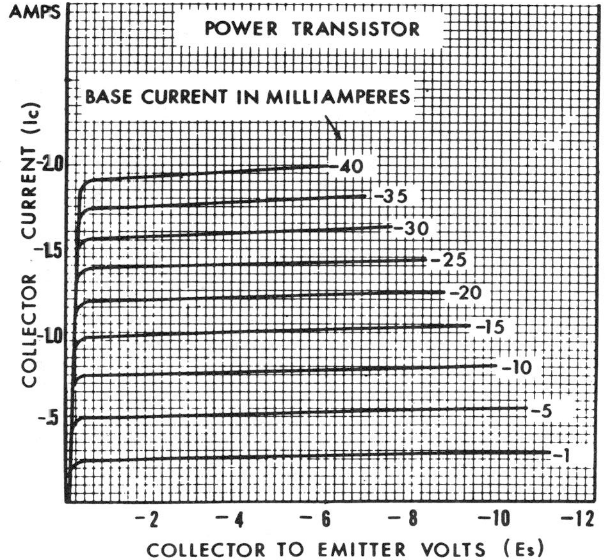

FIG. 15A

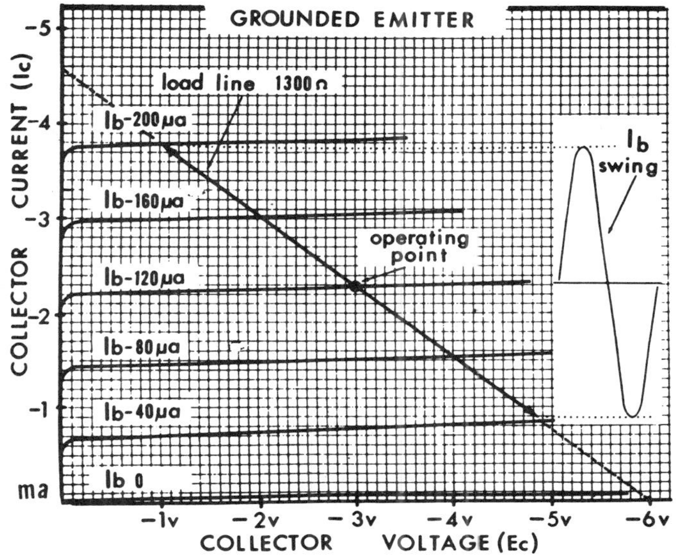

FAMILY OF CURVE TRACES

Fig. 15A is a family of curves showing increases in collector current for successive increases in base current. Notice that a small increase in base current is followed by a large increase of collector current. This conforms with the gain or "beta" of the transistor under test. The grounded-emitter curves show the operating base bias for this family.

16

IC Mixer Stage Tests

Diagram A in Fig. 16 is a schematic of a two-transistor package containing a PNP directly coupled to a NPN similar to the Darlington circuit. The overall benefit of this configuration is greater current amplification. Compare Oscillogram 2 with Oscillogram 1, the average for a single stage. Oscillo-

INTEGRATED CIRCUITS

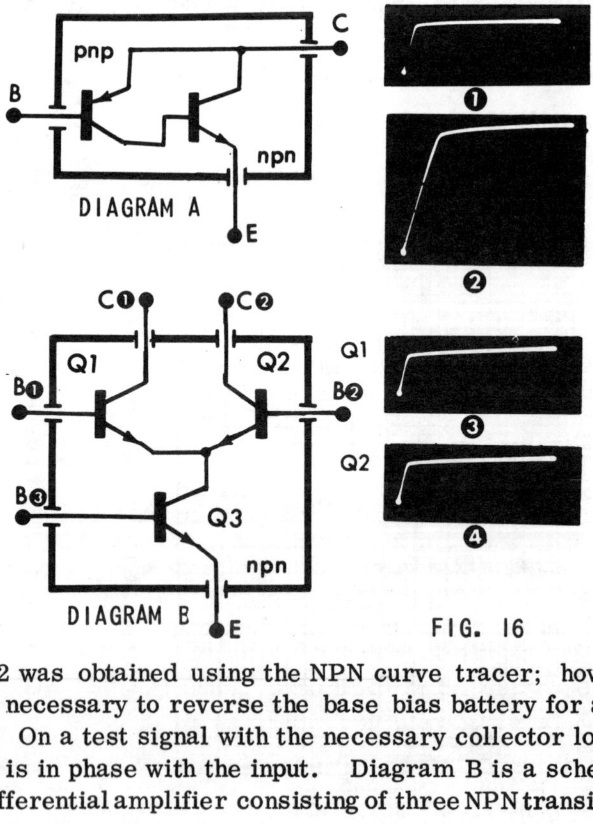

FIG. 16

gram 2 was obtained using the NPN curve tracer; however, it was necessary to reverse the base bias battery for a PNP input. On a test signal with the necessary collector load the output is in phase with the input. Diagram B is a schematic of a differential amplifier consisting of three NPN transistors.

Q1 and Q2 connected in differential amplifier configuration; Q3 is connected in series with the common-cathode connection. This stage is used for its constant-current characteristic and is referred to as a "heat sink." When a circuit is built around this package the base of Q3 is adjusted for a fixed base bias. This unit may be tested with the NPN curve circuit in Fig. 16. Two tests are required: one test for Q1 and another for Q2. When switched on the emitter of Q3 is common to both Q1 and Q2. A 1.5-volt battery and a 150K resistor were connected in series and placed between terminals B3 and E; positive to B3 and negative to E. The package offers many types of circuit arrangements—differential amplifer, multivibrator, mixer for superheterodyne front ends, etc.

FIG. 16A

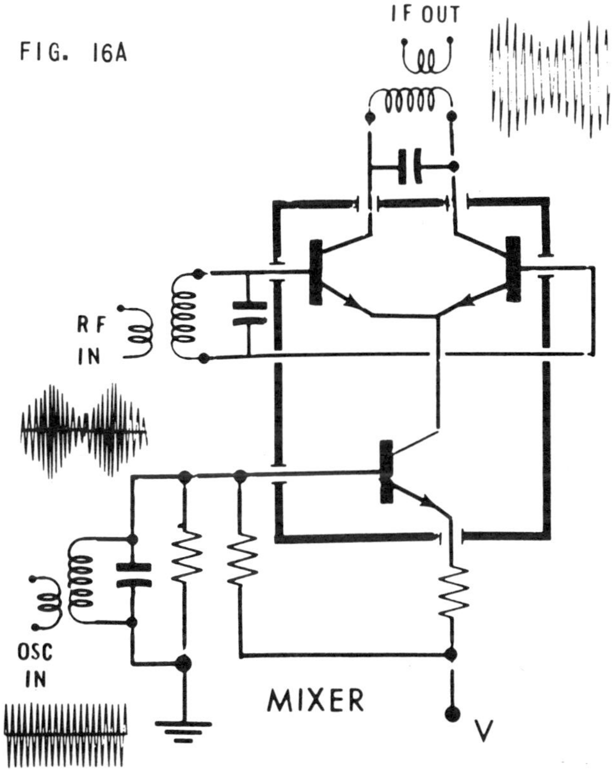

One of the many uses of the integrated differential amplifier is illustrated in Fig. 16A. The unit has a frequency range from DC to 250 MHz, ideal for superheterodyne communications receivers.

17

Measuring Collector Current and Voltage

Designed primarily for evaluation, the circuit illustrated in Fig. 17 produces a collector current and voltage curve shown in Diagram A.

Procedure

Equipment Required:

Oscilloscope
Curve tracer circuit

Connect the test circuit as shown in Fig. 17 and adjust vertical sensitivity to 1v/cm. Assume that the center of the curve is 3 centimeters above the zero reference, indicating 3 volts. This voltage is due to the collector current flowing through the 100-ohm resistor. Therefore:

$$I_C = \frac{E}{R} = \frac{3}{100} = 30 \text{ milliamperes}$$

Adjust the horizontal sweep control to 1v/cm and adjust the sweep to 6 centimeters. This will plot the collector voltage against the collector current. Current can be measured when the vertical sensitivity is calibrated and the resistor value (across which the voltage drop appears) is known.

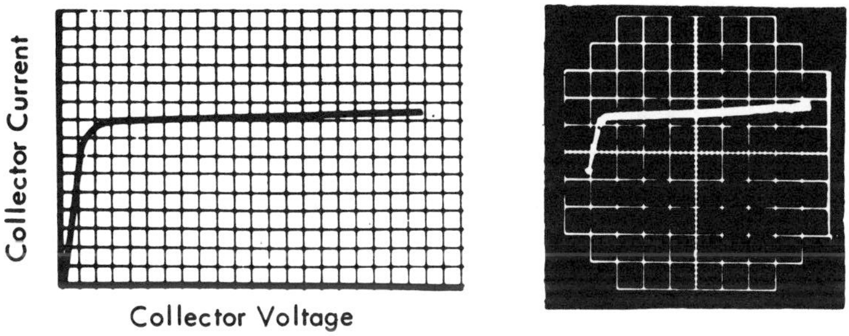

DIAGRAM A

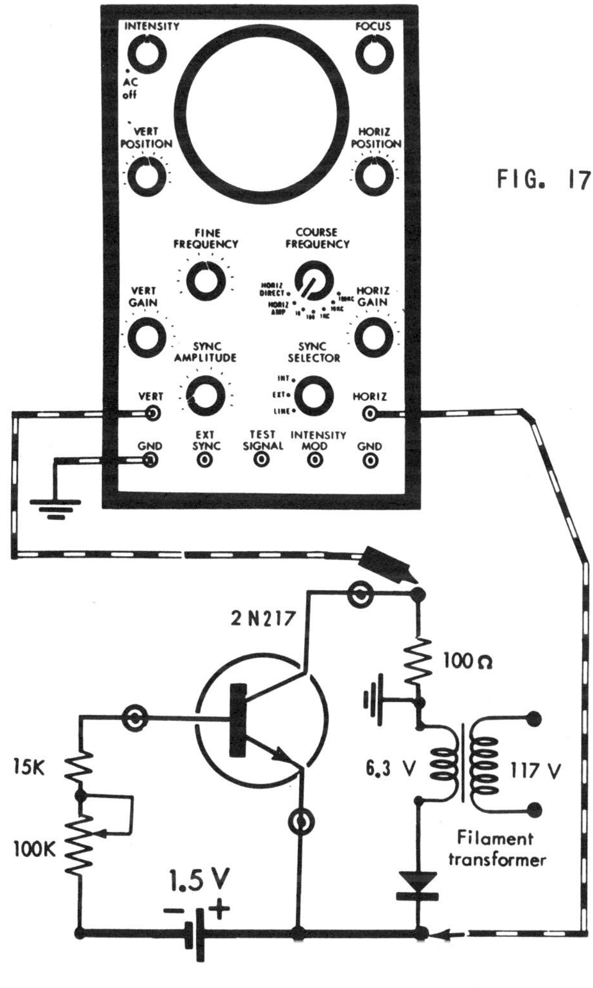

FIG. 17

18

Checking Class-A Transistor Amplifiers

The undistorted output of a Class A amplifier is dependent upon the operating base bias current. Assume a PNP stage where the bias current is too high. The positive peaks will distort before the negative peaks, and if the bias current is too low the negative peaks will distort before the positive peaks. In an NPN amplifier the polarity is reversed, of course. A simple check may be conducted to determine whether the bias current is too high or too low.

Procedure

Equipment Required:

Oscilloscope
Sine-wave generator (audio)

Connect the equipment as shown in Fig. 18. Tune the signal generator to 500 Hz and adjust the oscilloscope sweep rate to display 3 cycles. Adjust the fine frequency control to display the three cycles—three positive peaks and two negative peaks as illustrated in Diagram A. Gradually increase the generator volume control until a slight distortion is observed. If both positive and negative peaks distort at the same time, the base bias is correct and the input signal is at the threshold of distortion. A further increase will cause the peaks to flatten (the flat tops should be equal in duration). Oscillograms 3, 4, and 5 indicate incorrect bias; therefore, the base current should be reduced slightly

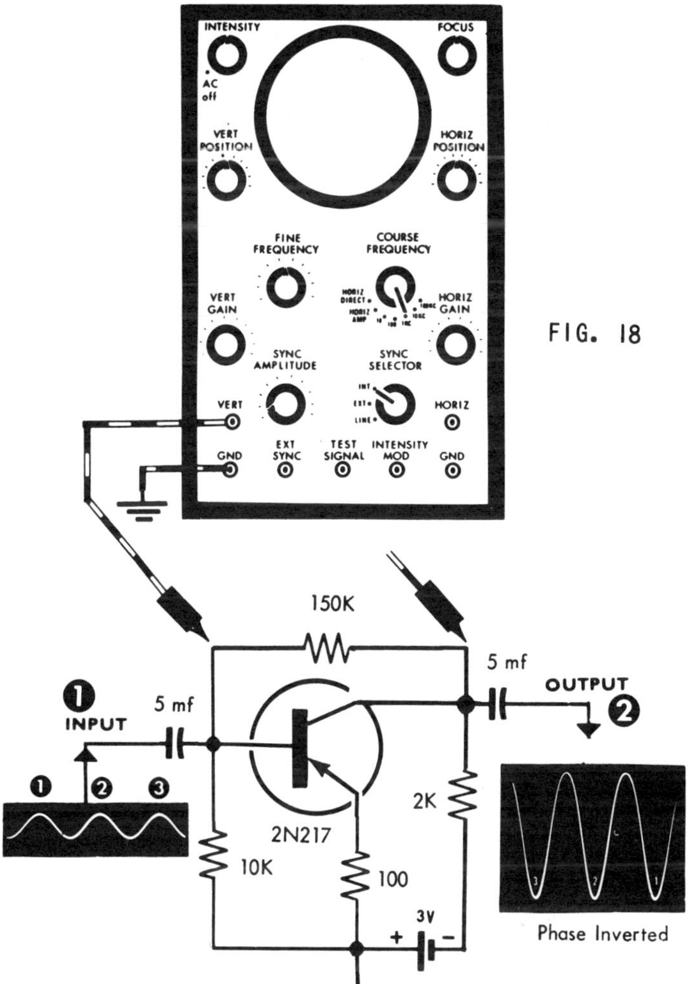

FIG. 18

DIAGRAM A

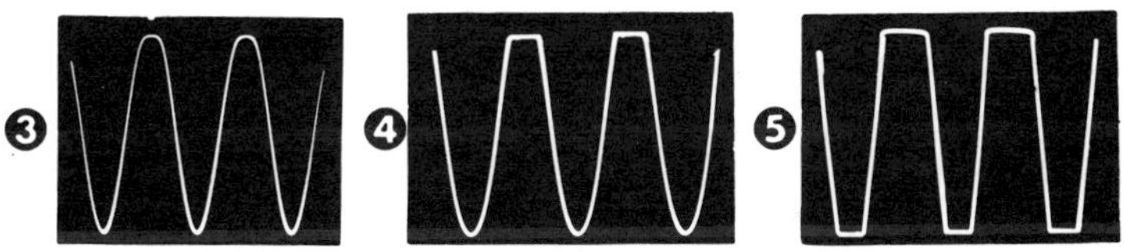

19

Checking Class-B Transistor Amplifiers

Class B push-pull amplifiers are biased just above cutoff. This type of circuit is very popular in transistorized audio amplifiers because of its low drain during standby periods. However, due to poor collector current linearity at low signal levels, there is a pronounced distortion as the signal swings through zero, alternating from one stage to the other. To prevent this complete cutoff of collector current a small bias is applied.

waveform across voice coil

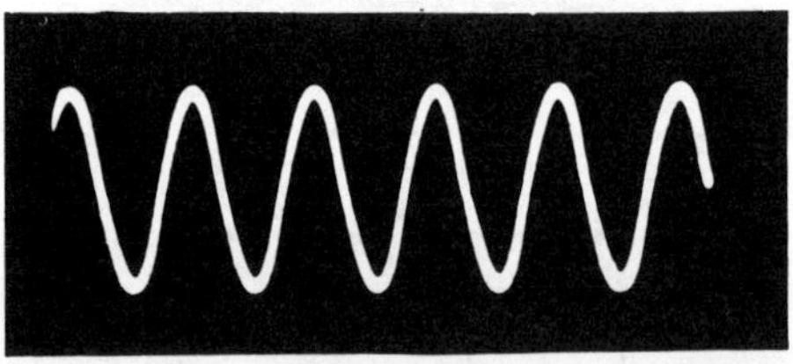

correct bias

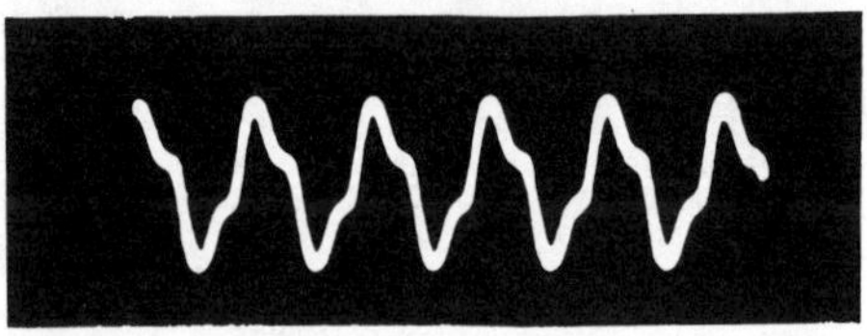

incorrect bias

Procedure

Equipment Required:

Oscilloscope
Sine-wave generator

Connect the audio signal generator to the input of the push-pull driver, or the primary of the input transformer if used.

See Fig. 19. Tune the generator to 1 kHz and connect the oscilloscope vertical input across the voice coil. Adjust both generator and oscilloscope to display a few cycles. The sine waves displayed should be free from distortion (see Oscillogram 1). Now, short out the bias resistor with a jumper wire and note distortion. Keep the signal generator input low to prevent distortion.

FIG. 19

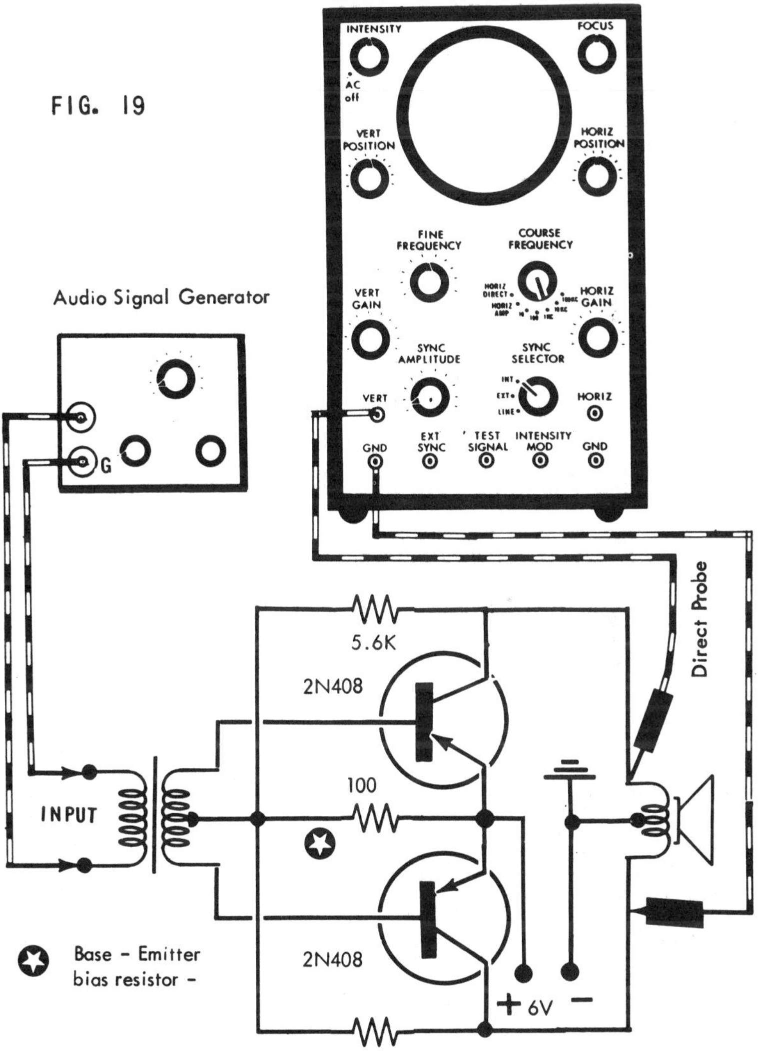

TYPICAL CLASS "B" AMPLIFIER

20

Stereo Amplifier Checks

The test setup in Fig. 20 shows a method for checking stereo channels. (With a dual trace oscilloscope or an electronic switch the two channels may be checked simultaneously).

Procedure

Equipment Required:

Oscilloscope
Square-wave generator

Check the square-wave generator output directly with the oscilloscope before making this test. Then, connect the output of the square-wave generator (at low volume) to the input of one channel, and connect the oscilloscope vertical input across the voice coil of that channel. Bass and treble controls must be set the same on both channels; i.e., maximum bass,

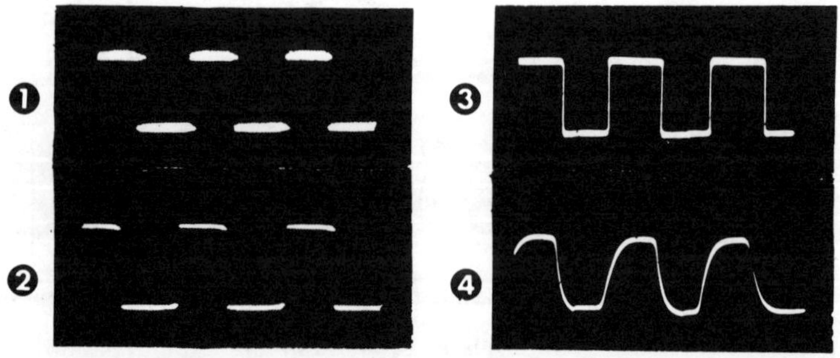

1. Hash on top and bottom indicates noise
2. Normal response
3. Noisy transistor
4. Effect caused by reduced treble control setting

maximum treble, and turn down the loudness control. Now observe the square-wave pattern at the output. If a fairly good square wave appears, then test the second channel and compare the two output waveforms. (The author ran this test with a dual trace oscilloscope and produced the oscillograms shown). It is advisable when conducting this test to check both bass and treble controls. Usually, a strong base will show ringing at top and bottom of square wave.

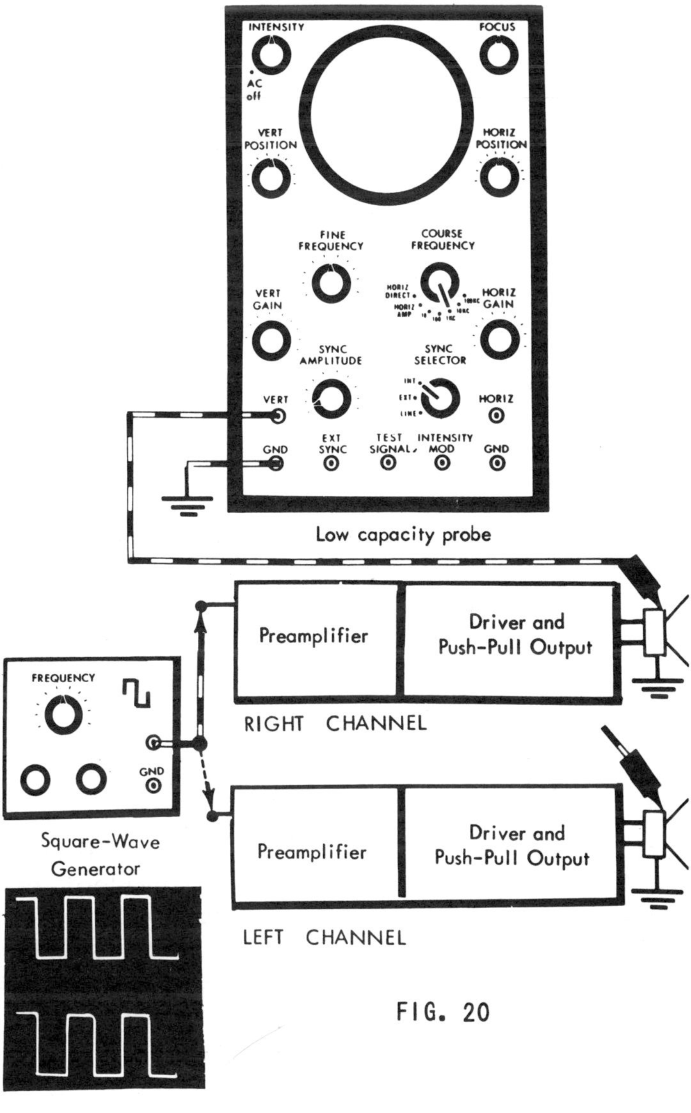

FIG. 20

21 Testing Audio Bypass Capacitors

Primarily, bypass capacitors are intended to remove AC signals from DC circuits. Therefore, an open or intermittent bypass capacitor can be quickly located with an oscilloscope.

Procedure

Equipment Required:

Oscilloscope
Well-shielded direct probe
Audio signal generator

A typical audio amplifier circuit is illustrated in Fig. 21. Connect the signal generator to position 1 and apply a 1000-Hz signal at very low volume. Check the signal at this position with the oscilloscope. Connect the probe to position 2 (output) and notice reasonable gain. Connect the probe to position 3 (cathode bypass capacitor) and turn up the vertical gain control to near maximum. If an appreciable audio signal is present at this point check the connections or replace the capacitor. A small amount of signal can be negative feedback and should not exceed more than 10% of the plate amplitude. Finally, connect the probe to position 4 (screen grid bypass capacitor). At this point there should be no evidence of signal present. In intermittent cases the capacitor under test should be probed with a wooden stick for loose connections or a defect.

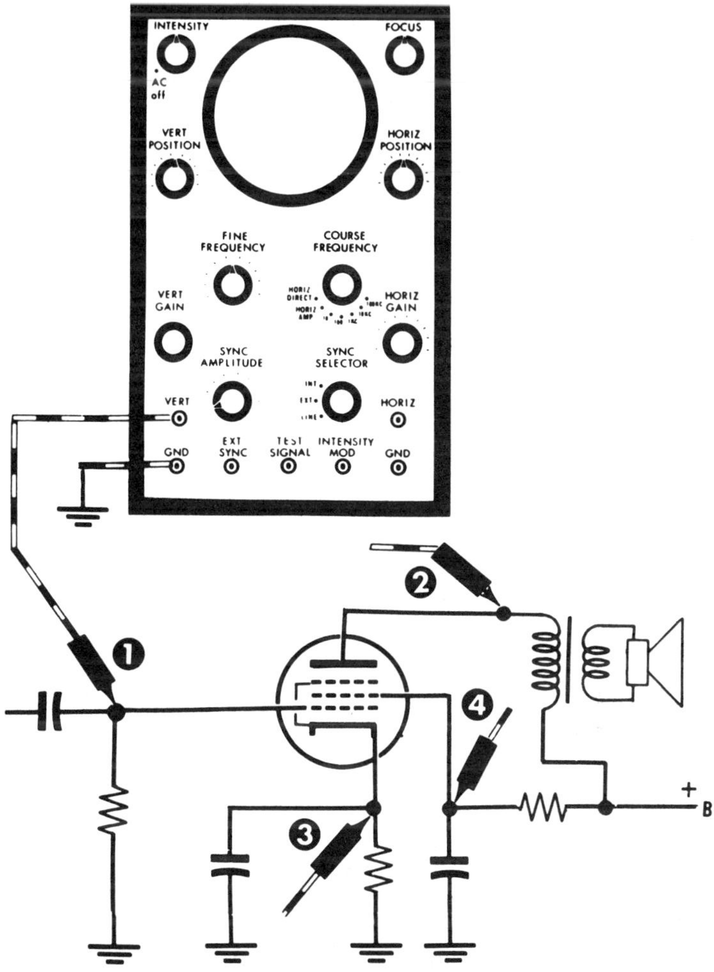
INTENSITY
FOCUS
AC
off
VERT
POSITION
HORIZ
POSITION
FINE
FREQUENCY
COURSE
FREQUENCY
VERT
GAIN
HORIZ
GAIN
SYNC
AMPLITUDE
SYNC
SELECTOR
VERT
HORIZ
GND
EXT
SYNC
TEST
SIGNAL
INTENSITY
MOD
GND
1
2
3
4
+
B

FIG. 21

22

Checking the Oscilloscope for Phase Shift

When measuring phase shift it is necessary to equalize vertical and horizontal traces, since the oscilloscope is being used as a vectorscope. This produces a diagonal line indicating zero phase shift between the vertical and horizontal amplifiers.

Procedure

Equipment Required:

Oscilloscope
Filament transformer

Connect the test circuit as shown in Fig. 22. Turn the vertical gain control to zero, turn up the horizontal gain, and adjust the horizontal trace for three inches or so many divisions. Remove the horizontal input lead and turn up the vertical gain control to three inches or so many divisions. Do not touch the controls until test is finished. Reconnect lead to the horizontal input and observe a diagonal trace, an indication that both amplifiers are in phase. If the trace is elliptical it indicates the presence of a phase shift (see Oscillograms 1, 2, and 3). The amount of phase shift is proportional to the width of the ellipse. On DC oscilloscopes a phase shift is less likely because the amplifiers are direct coupled.

❶

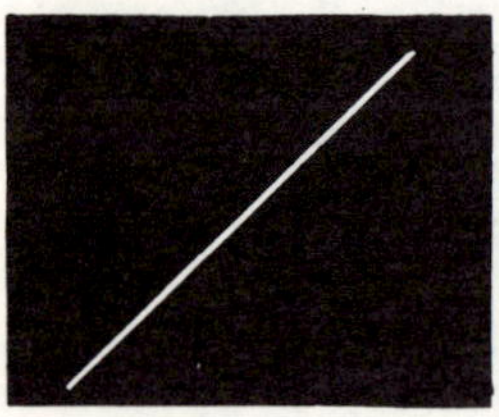

no phase shift

❷

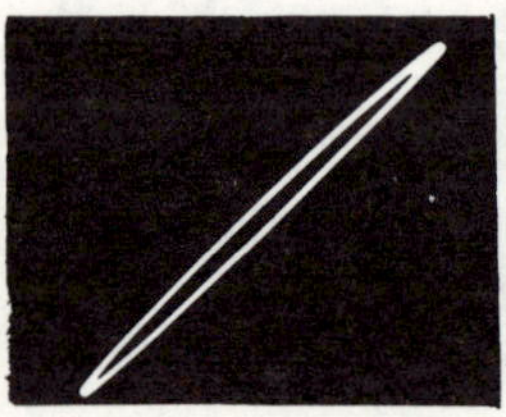

slight phase shift

❸

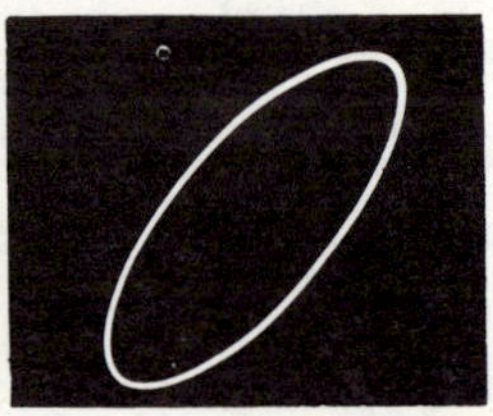

severe phase shift

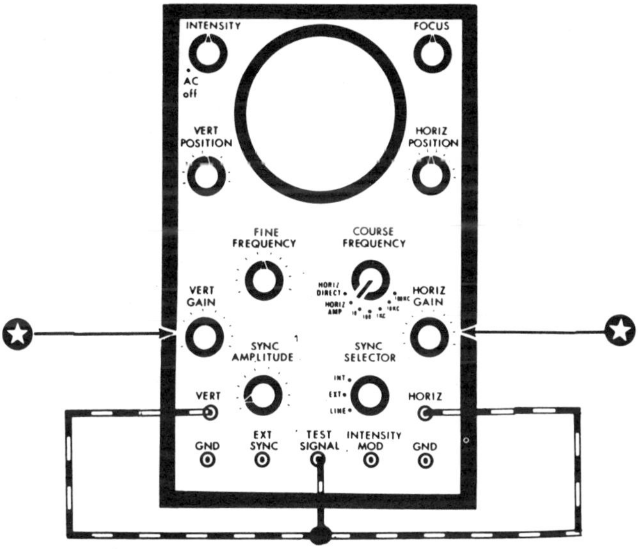

FIG. 22

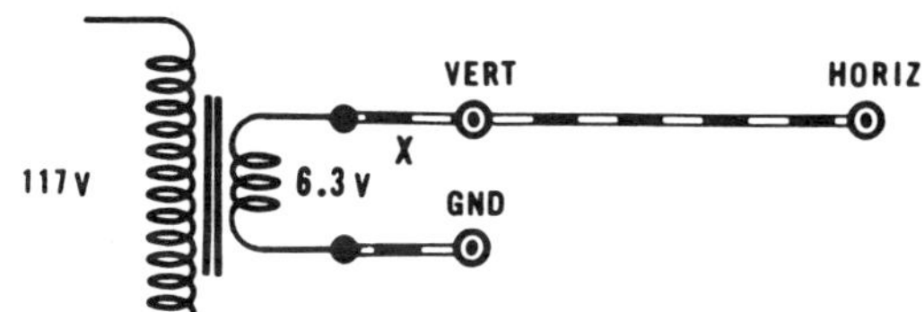

ALTERNATE HOOK—UP

EXAMPLE

❹

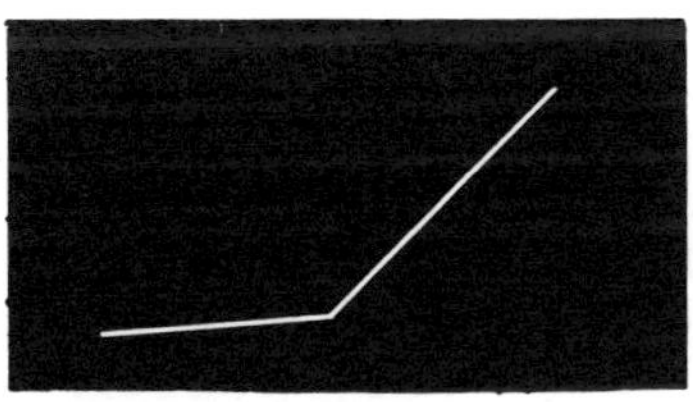

Indicates good diode.

❺

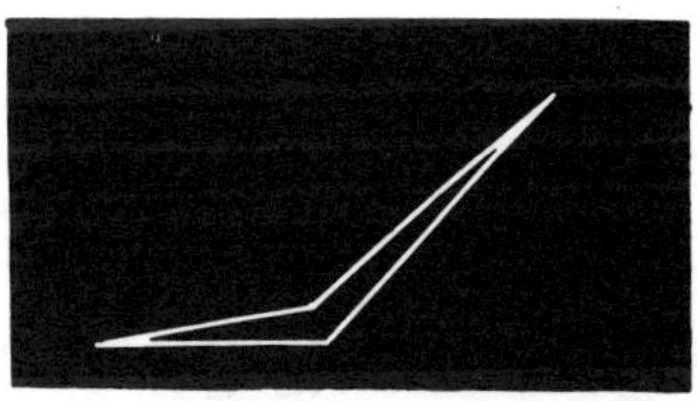

Dual trace indicates phase shift between H and V amplifiers.

23

Checking Detector Diodes

The test circuit shown in Fig. 23 provides a fullwave low-potential test for small diodes used in radio and television detector circuits. The voltage drop across R1 drives the diode and simulates a small AC signal. Diode current flows through R2, developing a voltage drop which is applied to the vertical input of the oscilloscope. The horizontal input is connected to the supply voltage as illustrated so that the curve is synchronized with the applied voltage.

Procedure

Equipment Required:

Oscilloscope
Test circuit

Turn both vertical and horizontal gain controls counterclockwise and adjust intensity and focus controls to obtain a small

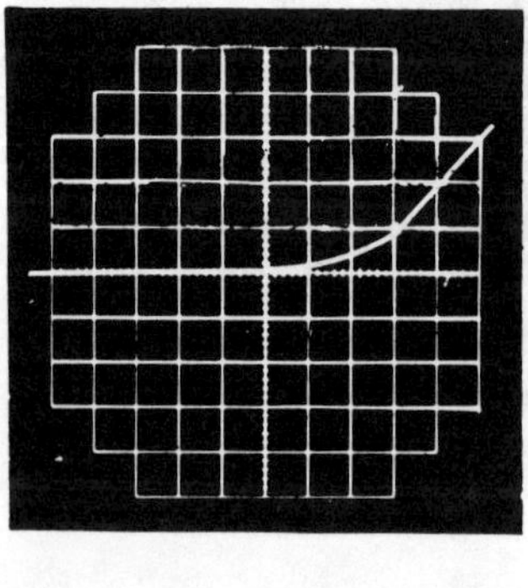

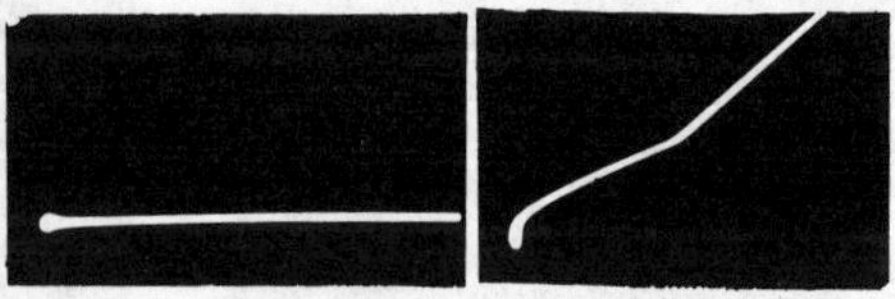

A—Shows a high forward to reverse current and indicates a good diode.

B—Indicates an open diode, however in this case check the test circuit connection before condemning the diode.

C—Indicates a partially or shorted diode.

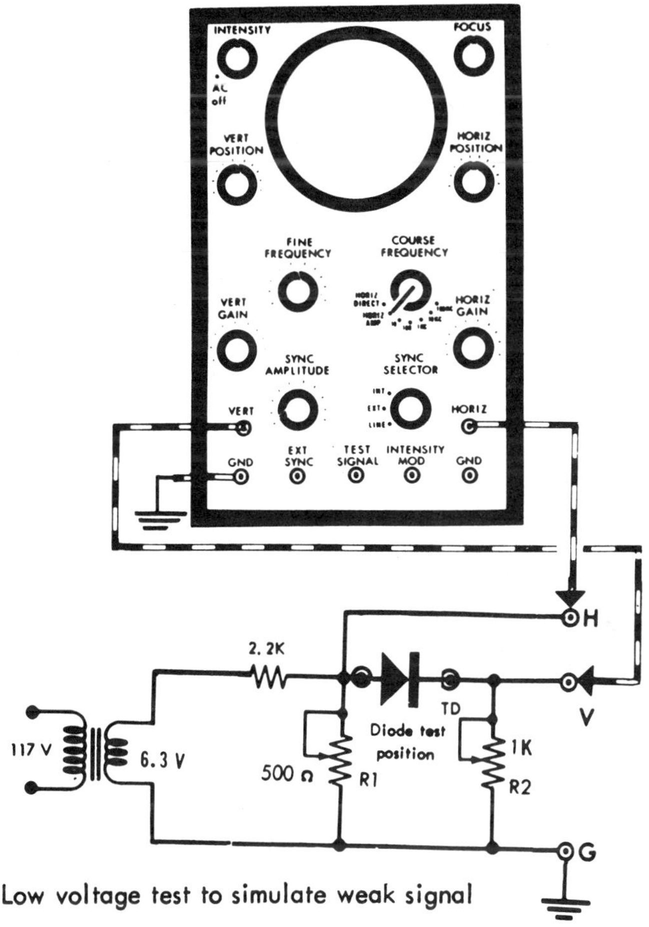

Low voltage test to simulate weak signal

FIG. 23

spot. Adjust the positioning controls until the spot is dead center on the graticule. Turn up vertical and horizontal controls and note vertical deflection from center to right if diode is normal. Compare the curve with the oscillogram.

24 Power Rectifier Tests

The voltage-current characteristics of a power rectifier can be observed by using the test circuit shown in Fig. 24. In this test the diode current develops a voltage drop across R1 which is applied to the vertical input of the oscilloscope, where vertical deflection indicates diode conduction for a half wave of the applied cycle. The flat portion indicates reverse bias.

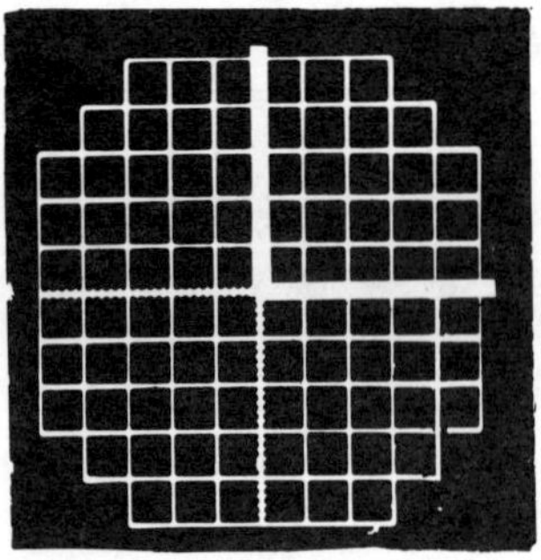

SILICON POWER RECTIFIER CURVE
Note abrupt switch from conducting to non- conducting..............

Procedure

Equipment Required:

Oscilloscope
Diode test circuit

Connect the oscilloscope vertical input across R1 and the horizontal input to the anode of the diode under test. The curve displayed should be compared with the oscillogram showing normal operation. A 6-watt 117-volt lamp is connected in

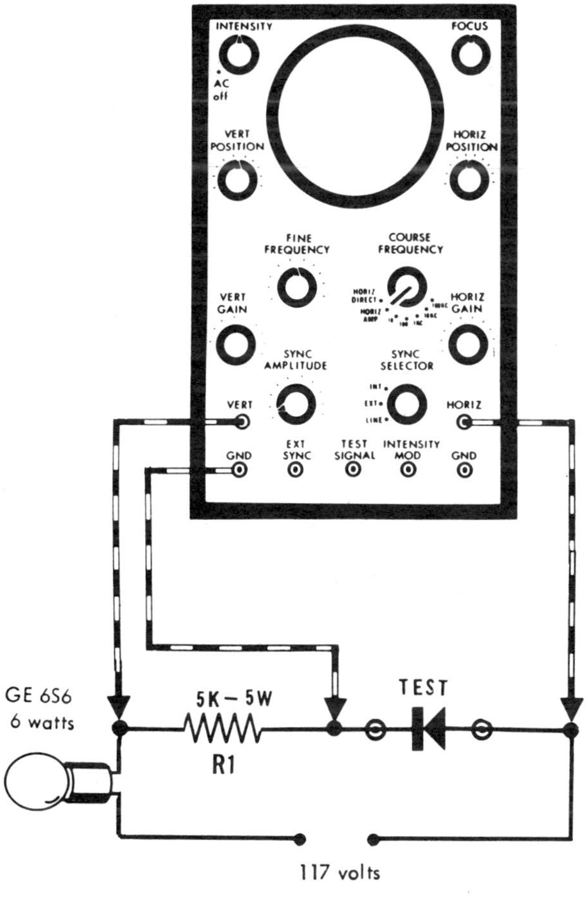

FIG. 24

series with the applied voltage for protection against an accidental short circuit. The oscillogram pattern was obtained from a silicon power rectifier intended for replacement in a color TV receiver.

25

Simple Phase Shift Circuit

The circuit shown in Fig. 25 is appropriate in many applications where phase control is necessary. It may be used as a phase corrector between two or more amplifiers.

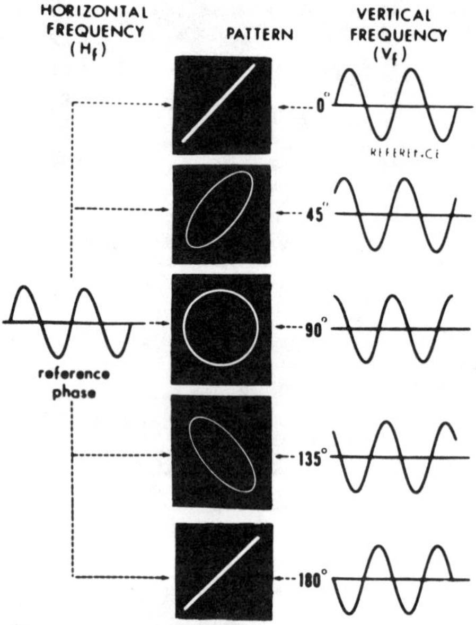

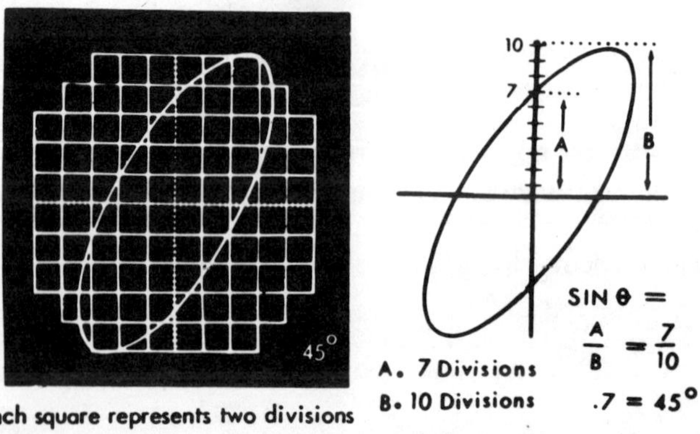

Each square represents two divisions

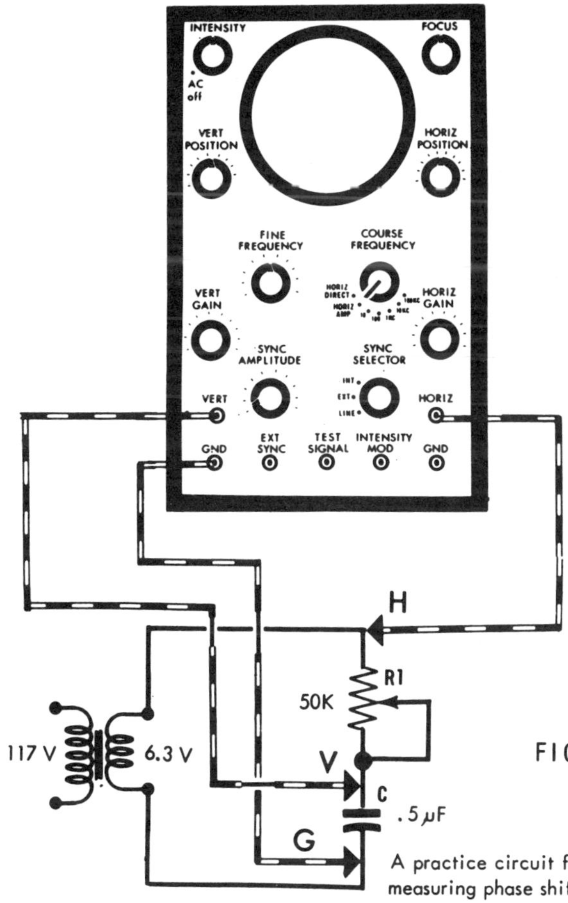

FIG. 25

A practice circuit for measuring phase shift

Procedure

Equipment Required:

Oscilloscope
Resistor—50K
Capacitor—0.5 mfd

Turn both vertical and horizontal gain controls fully counterclockwise and adjust both positioning controls until the spot is centered on the screen. Assuming the traces have been equalized, turn up the gain controls, then vary R1 and notice that the phase shift increases to 90°. See the oscillograms.

26

Comparing Square Wave Rise and Fall Time

A square-wave pattern may appear equal in rise and fall time, but this apparently ideal situation can be deceiving. In some laboratory tests it is necessary to know if any differences exist between the rise time of the leading edge and the fall time of the trailing edge.

Procedure

Equipment Required:

Oscilloscope, low-capacity probe
Square-wave generator
Trimmer capacitor and 10K resistor

Connect the output of a square-wave generator to the RC circuit shown in Fig. 26 and the oscilloscope vertical input across the 10K resistor. Tune the generator to about 100 Hz. Adjust the trimmer for sharp peaks and observe the amplitudes of both the positive and negative pulses. If the positive pulse is greater than the negative pulse, then the rise time is faster than the fall time. (See the oscillograms in Diagram A.) Due to the rapid charge of the very small capacitor the amplitude is a good indication of rise time. Also, the rapid capacitor discharge indicates fall time. The trigger pulses are actually very thin and just visible; therefore, to capture a photograph the intensity control had to be advanced. Sometimes the leading and trailing edges are not visible on the oscilloscope; however, by advancing the intensity control they can be made visible.

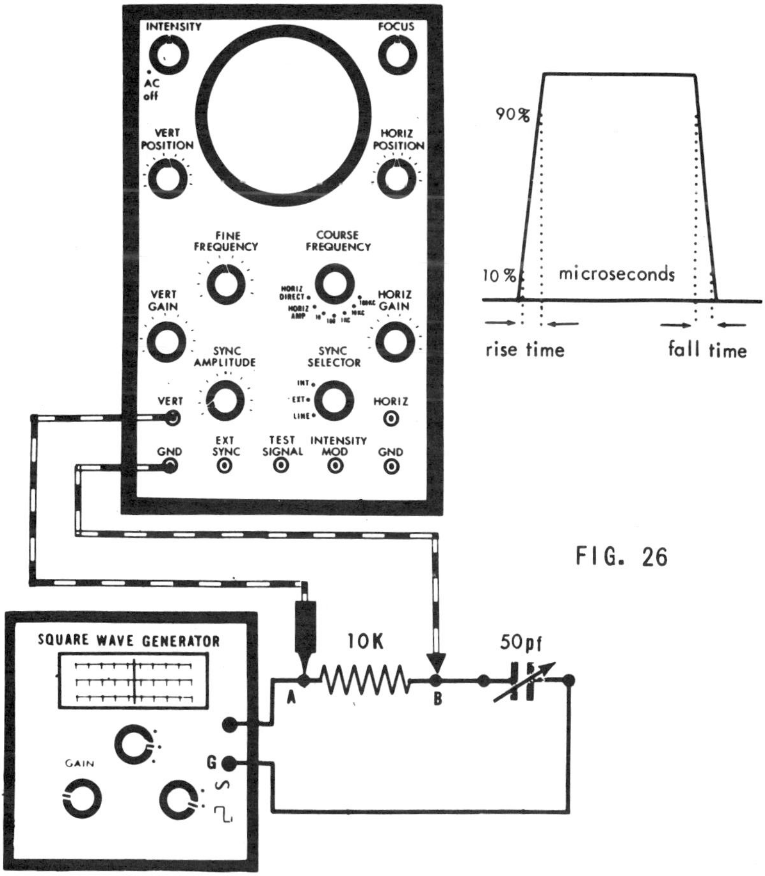

FIG. 26

DIAGRAM A

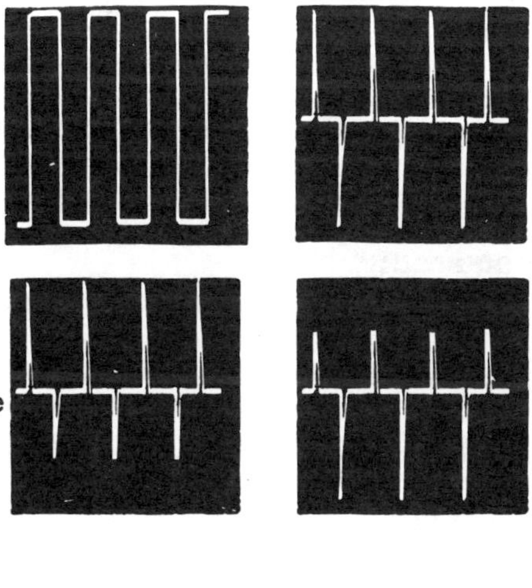

Square wave

Leading and trailing edges

Leading edge fast

Trailing edge fast

27

Transistorized Power Supply Waveforms

A transistorized power supply has no moving parts, it is compact, and has an operating efficiency as high as 90%. With these advantages and exceptional performance this type of power supply is gradually replacing the vibrator and dynamotor systems.

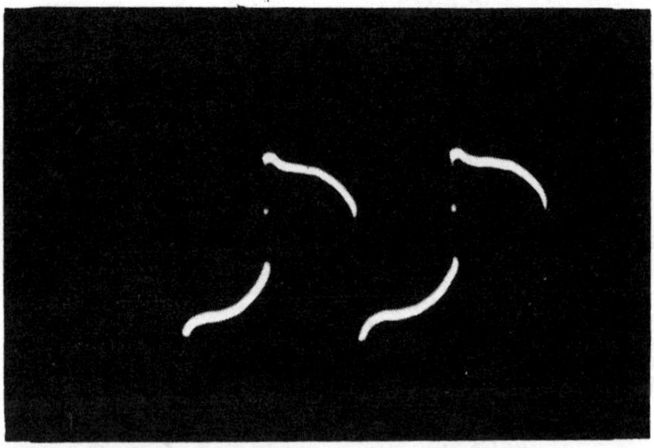

Procedure

Equipment Required:

Oscilloscope
Power supply components

The simple oscillator power supply shown in Fig. 27 uses two power transistors and is capable of converting a 1.5-volt battery source to 50 volts or higher, depending upon the transformer turns ratio. The extent of power output is governed by the specified wattage rating of the transistors and an efficient heat sink. On relatively large loads a storage battery would be required. The test procedure is shown in Fig. 27. Both collector and base waveforms must be identical for each stage; any departure from the normal waveform indicates a component malfunction.

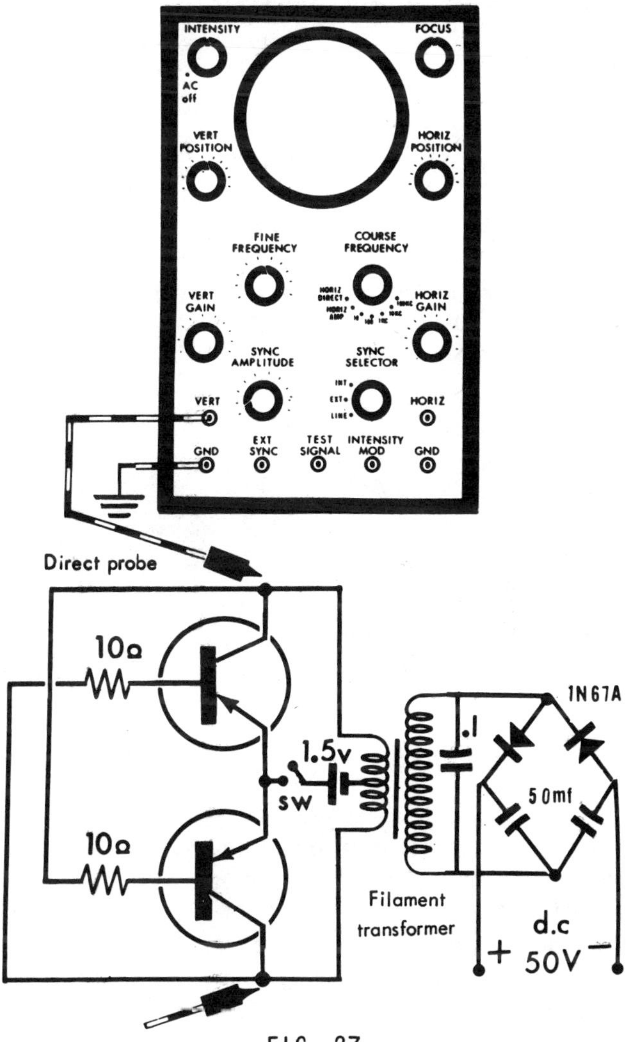

FIG. 27

28

Transistorized Power Supply with Separate Bias Coils

An improved version of the transistorized power supply is shown in Fig. 28. The operating efficiency is improved by using a separate base feedback winding, and the improvement is noticeable in the collector waveform shown in the oscillograms in Diagram A. The base winding stepdown ratio is 20-to-1, and 20-ohm base resistor R2 forms a voltage di-

Collector waveforms

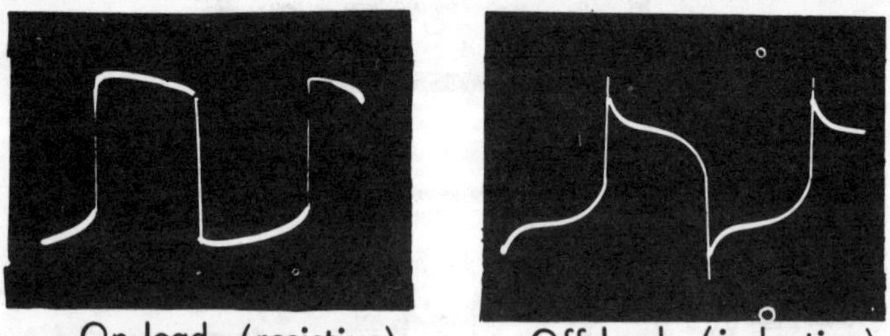

On load (resistive) Off-load (inductive)

DIAGRAM A

vider with R1, a control used to adjust the base current for normal operation while observing the collector waveform. The output voltage is 117 volts AC, and it may be rectified by a voltage doubler to provide 360 volts DC. This, of course, would limit the output current to a few milliamperes. The circuits discussed are the basis for heavier duty regulated power supplies now available on the market.

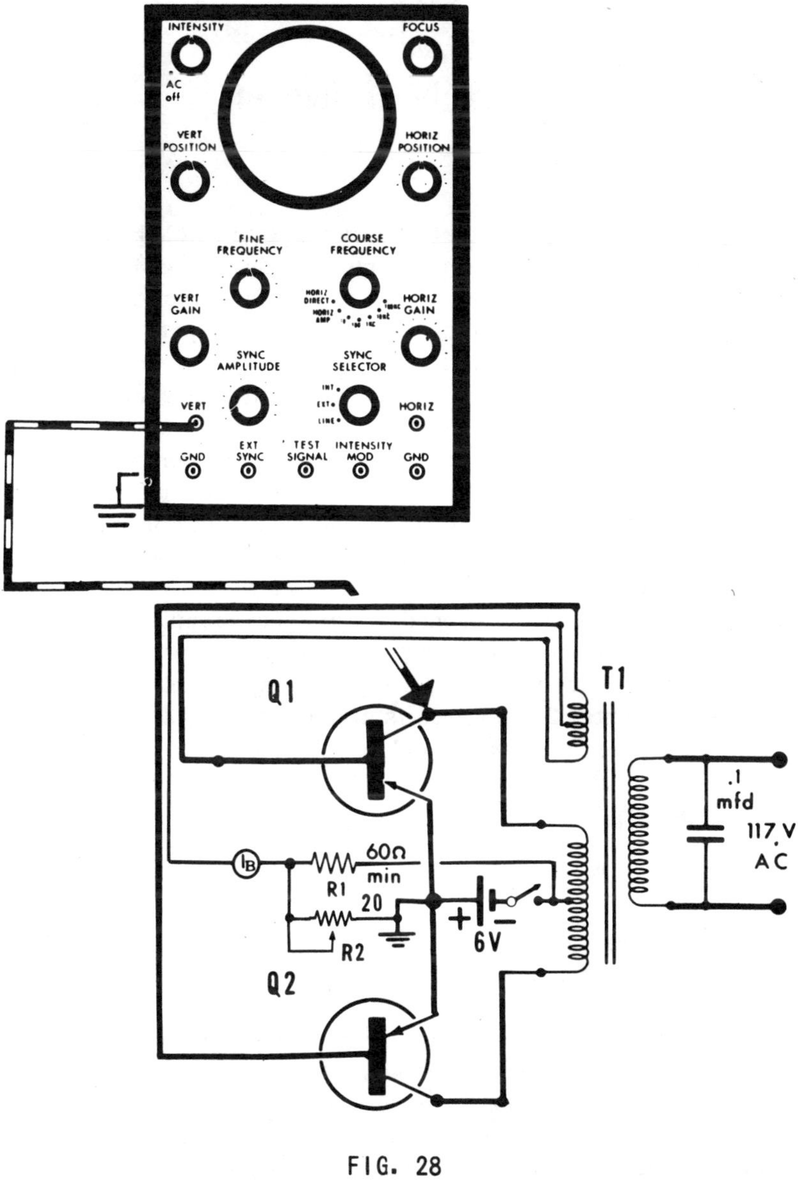
INTENSITY
FOCUS
AC
off
VERT
POSITION
HORIZ
POSITION
FINE
FREQUENCY
COURSE
FREQUENCY
VERT
GAIN
HORIZ
GAIN
SYNC
AMPLITUDE
SYNC
SELECTOR
VERT
HORIZ
GND
EXT
SYNC
TEST
SIGNAL
INTENSITY
MOD
GND
Q1
T1
.1
mfd
117 V
AC
60Ω
min
R1
20
R2
6V
Q2

FIG. 28

29

Intensity Modulation

The X and Y axes of a cathode-ray tube are well known, but a third axis used for intensity modulation and designated the Z axis has not been discussed too often, despite its many useful applications. Some oscilloscopes are provided with a panel terminal for intensity modulation. If a terminal is not provided, refer to the manufacturer's operating manual regarding this procedure. However, intensity modulation simply explained is actually a pulse applied to the control grid or cathode of a cathode-ray tube—a negative pulse to the control grid or a positive pulse to the cathode. It is similar to modulating the cathode-ray beam of a TV receiver with a video signal. (In a TV receiver all three axes, X, Y, and Z are active). In laboratory procedures a signal is applied to either element mentioned above to produce a pattern of dashed lines or dots that have a particular significance in timing and other measurements. The system is also used to blank out retrace during flyback time in TV and oscilloscopes.

Procedure

Equipment Required:

Oscilloscope
Square-wave generator

Tune the square-wave generator to a frequency several times higher than the sweep frequency. In order to produce precise time markers with decisive sharpness a square-wave or a series of pulses are required. However, in this procedure a 60-Hz sine wave is used. Connect the equipment as shown in Fig. 29 and adjust the oscilloscope to display a sine wave at 60 Hz. Gradually turn up the signal generator gain control

FIG. 29

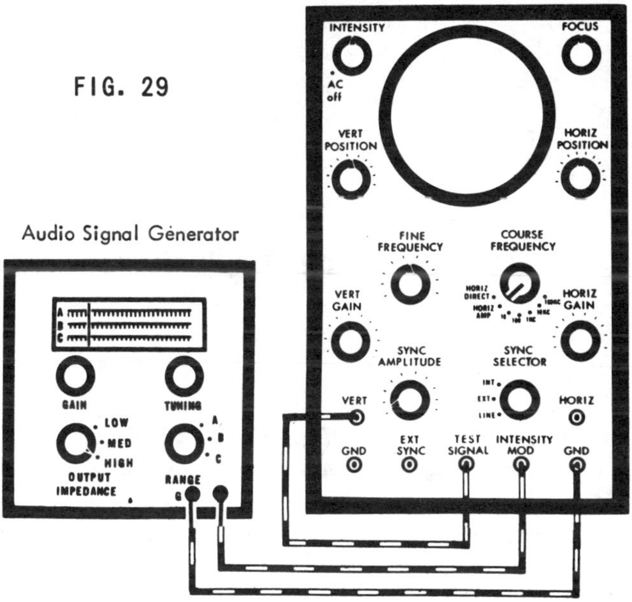

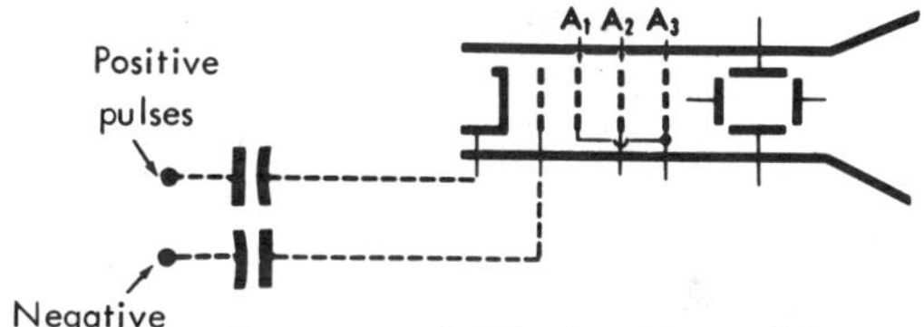

Alternate method for intensity modulation

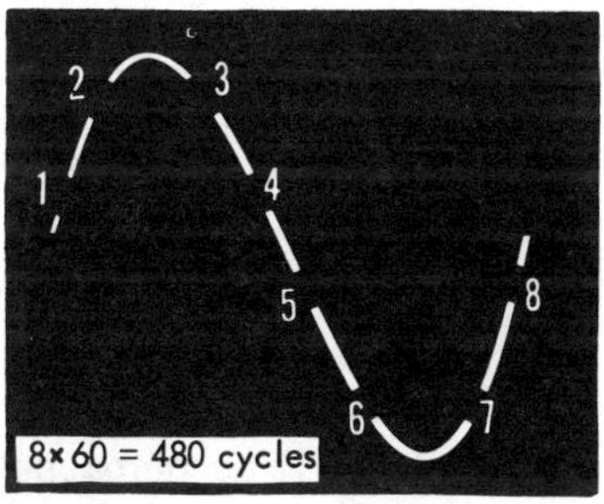

DIAGRAM A

until breaks begin to appear in the sine wave (see Diagram A). Adjust the intensity control slowly until the pattern shows a dashed line. Reduce the vertical gain control to zero to display a dashed time base.

30 Circular Trace Applications

The bridge circuit shown in Fig. 30 provides a circular trace which offers many applications such as checking the oscilloscope for astigmatism (defocussing of the cathode-ray beam in certain outer screen areas). The trace also is used in intensity modulation for measuring frequency (see Diagram A).

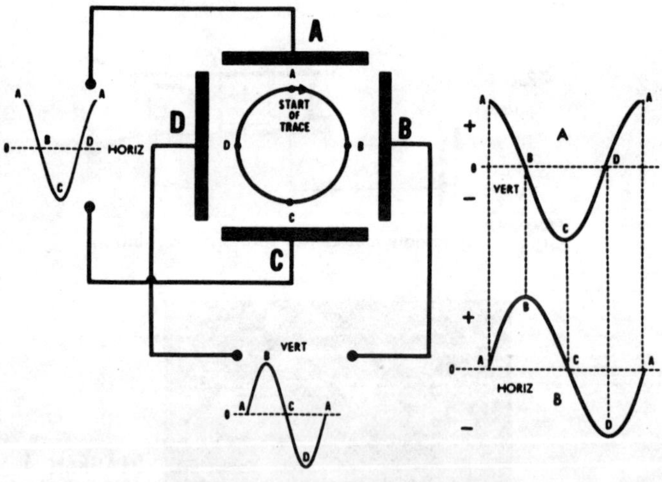

DIAGRAM A

It is used also in radar since it encompasses 360°. Actually the circuit converts a 360° sine wave into a 360° circular trace. Notice the effect on the trace when the sine wave is distorted. The diagram illustrates the trace of the cathode-ray beam when a sine wave is applied to the deflection plates of the oscilloscope.

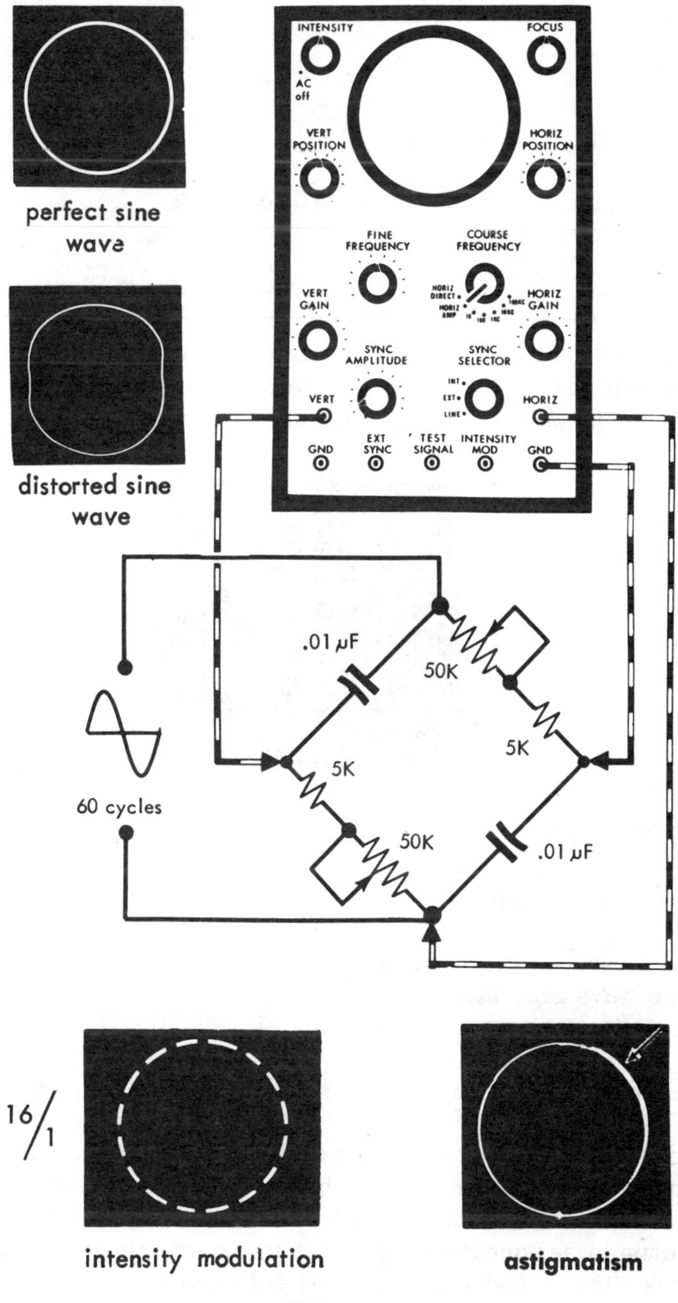
perfect sine wave
distorted sine wave
INTENSITY
FOCUS
AC off
VERT POSITION
HORIZ POSITION
FINE FREQUENCY
COURSE FREQUENCY
VERT GAIN
HORIZ GAIN
HORIZ DIRECT
HORIZ AMP
SYNC AMPLITUDE
SYNC SELECTOR
INT
EXT
LINE
VERT
HORIZ
GND
EXT SYNC
TEST SIGNAL
INTENSITY MOD
GND
.01 µF
50K
5K
5K
50K
.01 µF
60 cycles
16/1
intensity modulation
astigmatism

FIG. 30

31

Matching Capacitors

Sometimes it is necessary to match capacitors for critical applications. If a capacitor tester is not available, an oscilloscope will do a precise job of matching capacitors.

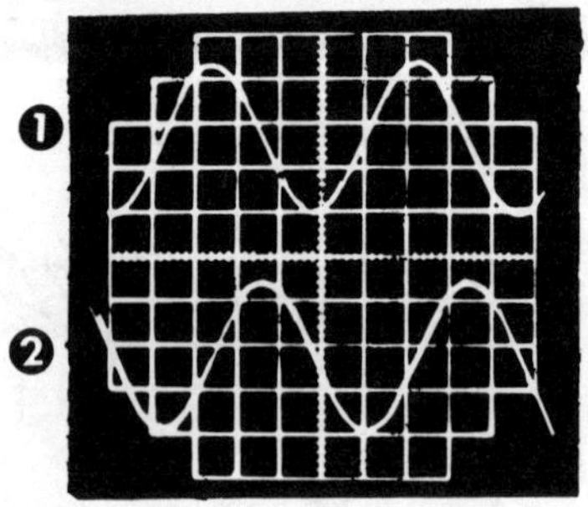

Procedure

Equipment Required:

Oscilloscope, direct probe
Sine-wave generator
Test circuit

Prepare the test circuit as shown in Fig. 31. Connect the ground terminal of the oscilloscope to the probe and the vertical input lead to the junction of the two capacitors. This is important to prevent pattern jitter. Connect a sine-wave generator to the input of the test circuit and tune it to 1 kHz (adjust sweep accordingly). Then probe with the ground lead at position 1 and position 2. If both capacitors are equal, the signal across each one will be the same amplitude. If unequal then the larger amplitude will indicate the smaller capacitor. The ratio of the two amplitudes in volts will be equal to the ratio of the two capacitors in mfd, pfd, or whatever the size of the capacitors under test.

FIG. 31

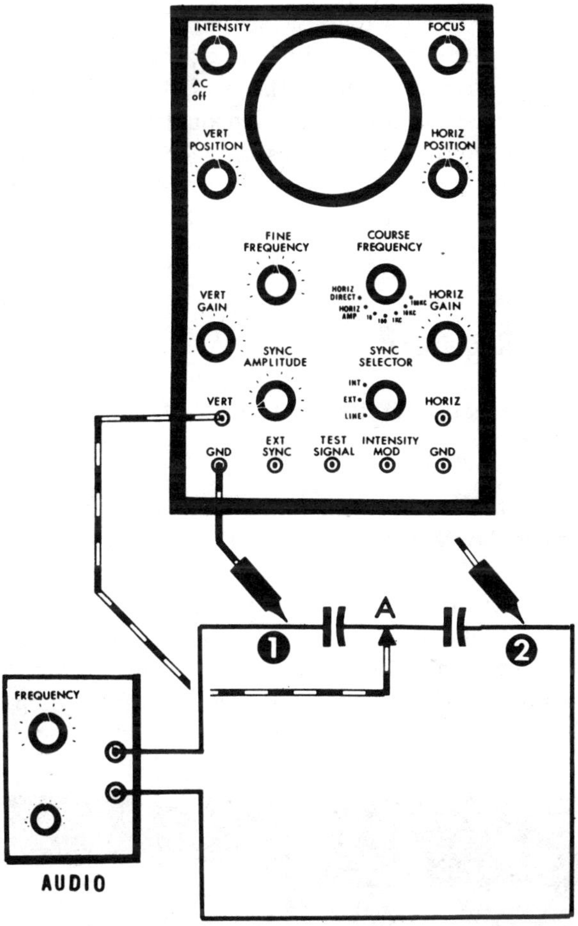

32

Checking Transistors for Noise

A square wave is an excellent means of checking noise in a transistor amplifier, since any noise will appear as "hash" riding on the flat portions of the wave—especially the flat portion that occurs during heavy conduction. A test was conducted by the author on a transistor known to be noisy. A simple amplifier circuit was constructed for quick insertion of suspected transistors.

Procedure

Equipment Required:

Oscilloscope—well shielded probe
Test circuit

Construct a test circuit as shown in Fig. 32 and insert the transistor to be tested. Connect the oscilloscope vertical input to the collector, leaving both input and output circuits open as illustrated. Adjust the vertical gain for maximum sensitivity. Some hash is permissible since the circuit and oscilloscope are operating wide open. Oscillogram 2 shows a very noisy transistor from a stereo circuit. Compare this

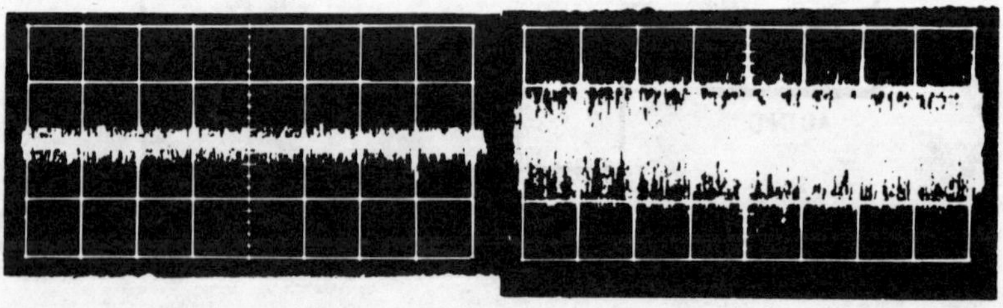

noise pattern with a comparable transistor that was the replacement (Oscillogram 1). To prevent misleading results, remove the circuit from the oscilloscope input terminal and observe "quiet," or noise if oscilloscope is defective.

FIG. 32

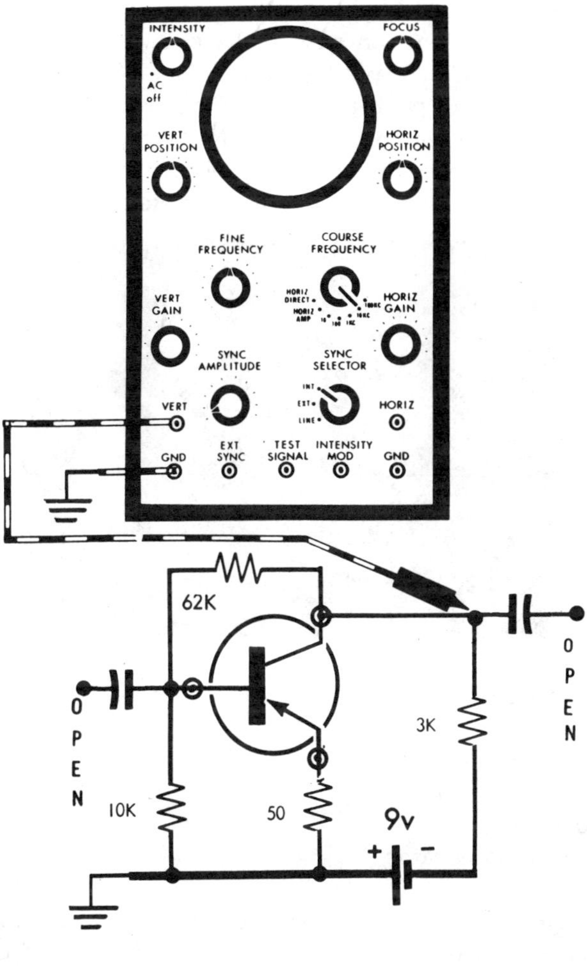

33

Paraphase Amplifier

Push-pull amplifiers require two inputs of opposite phase which are usually obtained from a push-pull input transformer or in conjunction with a phase inverter stage. The paraphase amplifier circuit shown in Fig. 33 employs a split load that supplies a push-pull output from a single-ended input. (Notice this stage is a combined common-emitter and emitter-follower.) However, the two output signals must be equal and will require checking if distortion is present.

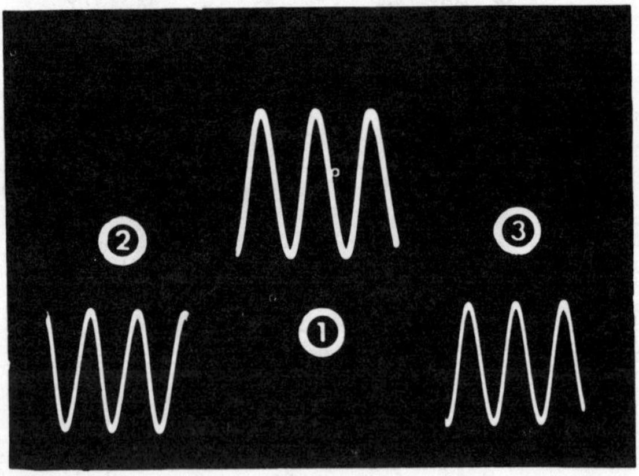

Procedure

Equipment required:

Oscilloscope
Audio signal generator

When load resistors R3 and R4 are equal, the stage will provide a nearly balanced output. An unbalanced condition can occur by a change in the value of either R3 or R4. For example, a change in the collector load resistor can only affect the collector output, but a change in the emitter load resistor can affect both outputs due to negative feedback. Connect the signal generator output to the audio amplifier. Check the output at the collector and compare this signal with the signal appearing at the emitter. They should balance within 5%. Oscillogram 1 is the input signal, Oscillogram 2 the collector output signal, and Oscillogram 3 the emitter output signal. Notice the output signals are less than unity gain, actually measured at 99% of the input.

FIG. 33

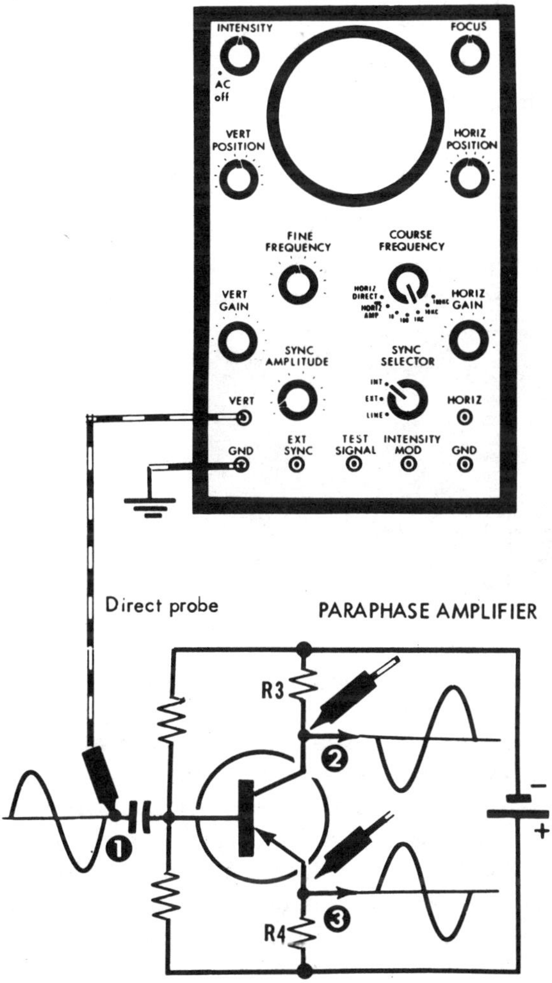

34 Measuring Audio Amplifier Gain

A convenient gain test is shown in Fig. 34.

Procedure

Equipment Required:

Oscilloscope
Audio signal generator

Connect the test equipment to the amplifier as shown in Fig. 34 and the oscilloscope probe to the signal generator output

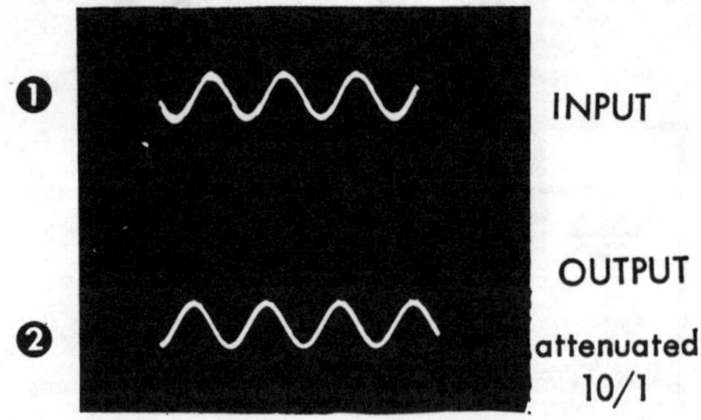

(position 1). Keep generator output low. Set the sweep frequency to display three cycles and record the sine wave amplitude in divisions. Switch the vertical gain attenuator to the 10-to-1 position (x10) and connect the probe to the output of the generator (position 2). If the output signal amplitude is the same as the input, the gain of the amplifier is 10. If it is twice the amplitude the gain is 20, and so on. For example, if the input was adjusted for one volt per division, then the output would be 10 volts per division. If the oscilloscope is not equipped with a decade attenuator, then a 10-to-1 attenuator probe should be used. The oscillograms were taken from a dual trace oscilloscope (top trace is the input and the bottom trace is the output). Notice the phase inversion. The amplifier was a one-tube affair with a gain of 10 to 1, built especially for the test.

FIG. 34

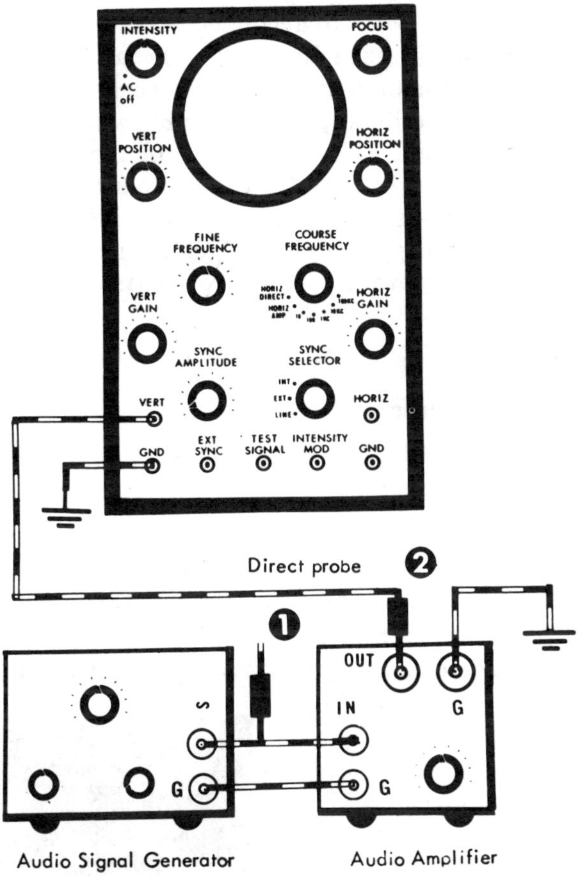

35

Dual Trace Using an Electronic Switch

The electronic switch is actually a square-wave generator consisting of a variable-frequency multivibrator which alternately gates two independent amplifiers on and off. The input of each amplifier is connected to a terminal on the panel and designated "A" and "B" to accommodate two signals. Each signal in turn is switched alternately through its respective amplifier to a common output, which is connected to the vertical input of an oscilloscope. The rapid switching rate permits the two signals to be observed simultaneously. The balance control is used to position the two signals on the oscilloscope screen—signal A is above signal B, for example. During the reversing process A moves down and B moves up, and the two may be overlapped for comparison.

Procedure

Equipment Required:

Oscilloscope
Electronic switch

The output of the electronic switch is connected to the vertical input of the oscilloscope, which is adjusted to display square waves. Turn the gain controls on the electronic switch to zero and on the oscilloscope to obtain two parallel traces (see

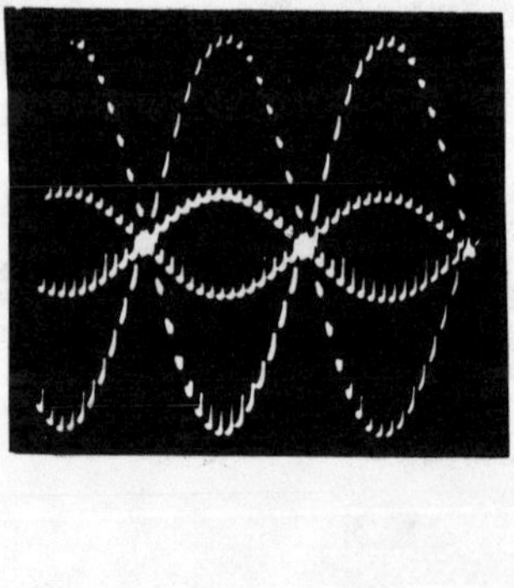

Horizontal sweep frequency twice the vertical frequency

Fig. 35). Then gradually turn up both gain controls (A and B) so that both signals may be evaluated. Since the output of the electronic switch contains the switching frequency it rules out the use "int sync"; therefore, switch the oscilloscope to "ext sync" and connect a lead from the "ext sync" post on the oscilloscope to either terminal A or B on the electronic switch.

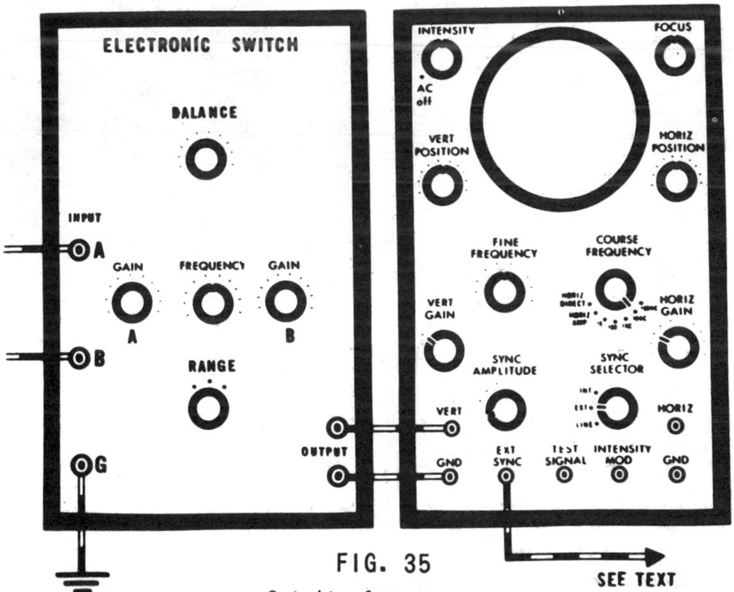

FIG. 35

Switching frequency

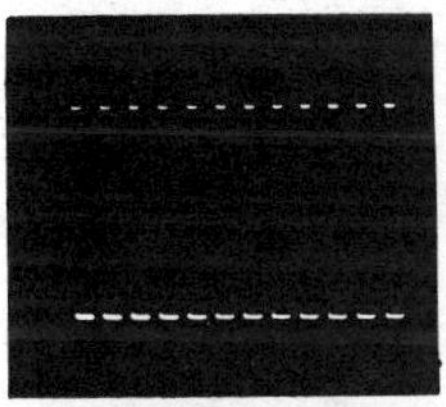

1 kilocycle

5 kilocycles

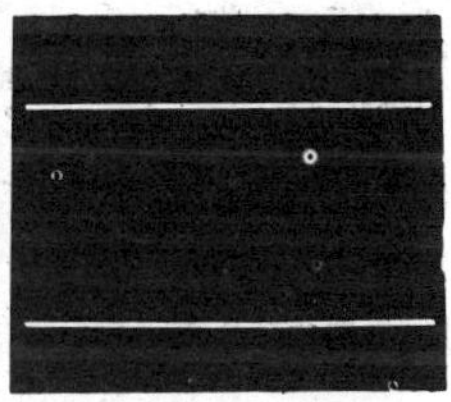

10 kilocycles

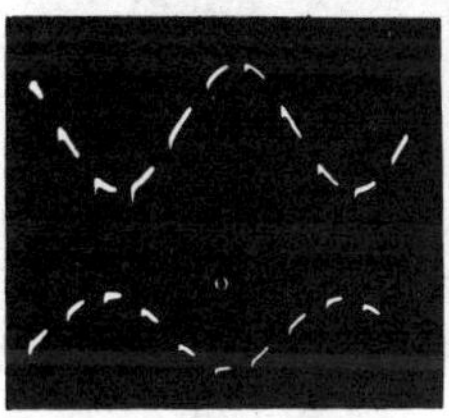

Input and output signals of amplifier shows poor gain

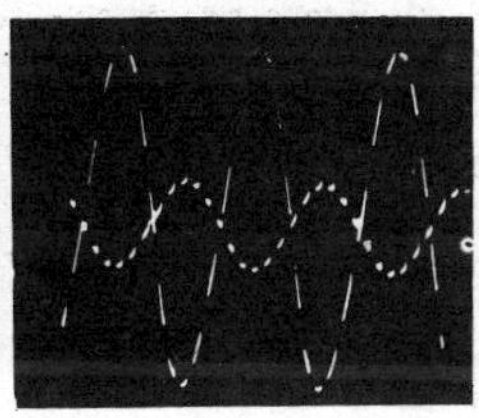

Input and output signals converged by balance control

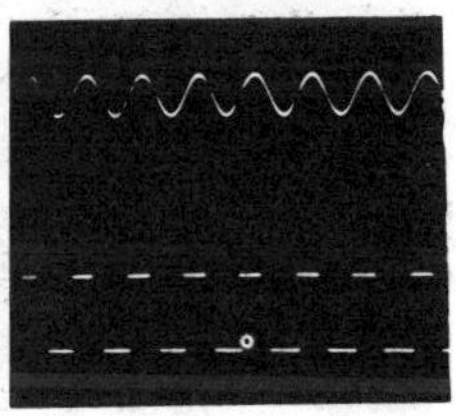

Schmitt Trigger Circuit
Input sine wave
Output square wave

36 Triple Trace Using Two Electronic Switches

There are occasions when it is advantageous to view three signals simultaneously. This may be achieved by using two electronic switches as shown in Fig. 36. The two signals present in the output of the first switch are fed as one to the input of the second switch, and the third signal is fed to the second input on the second switch. With additional electronic switches more signals can be applied. However, the number of signals permissible for simultaneous viewing is limited by the area of the screen. Signals to be observed must be of the same frequency, and the oscilloscope external sync should be connected to signal 1 at the electronic switch input.

TRIPLE TRACE OSCILLOGRAMS

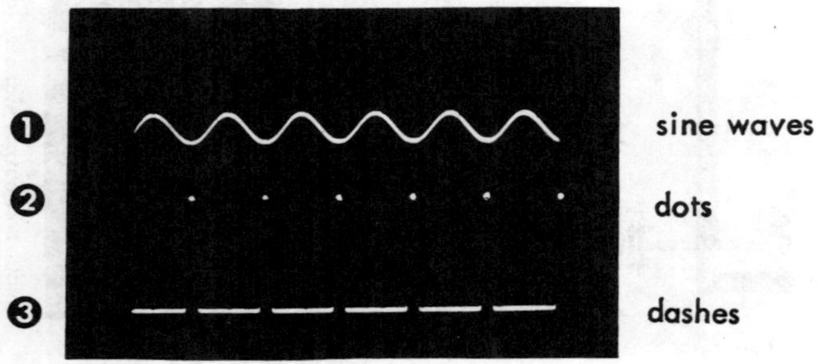

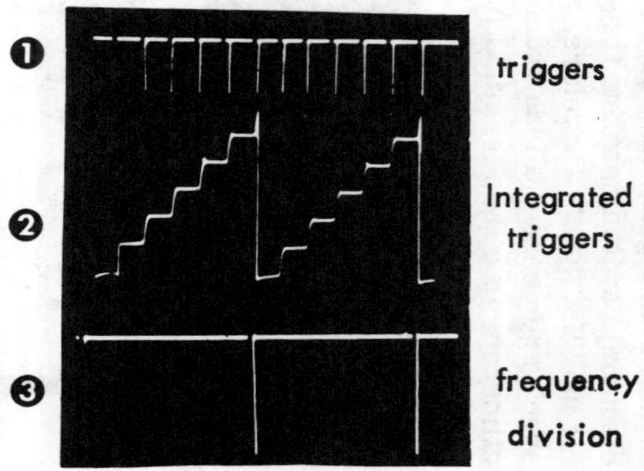

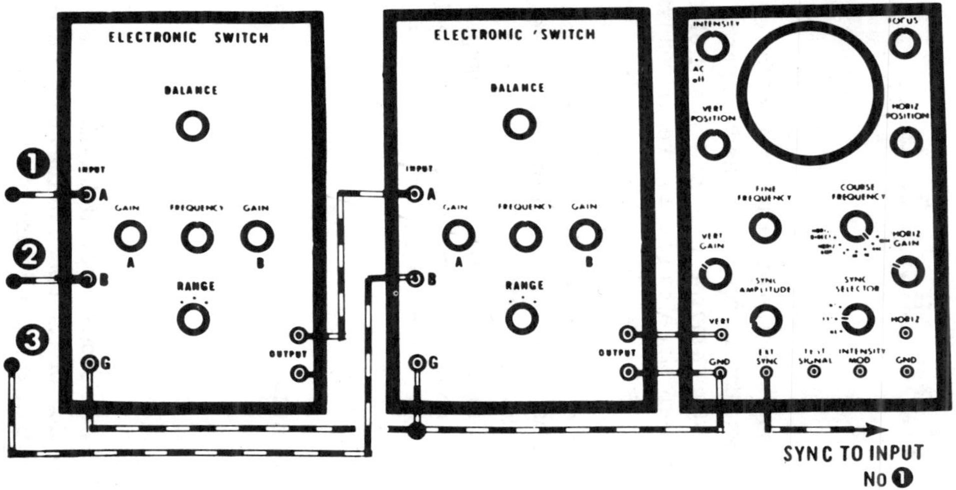

ILLUSTRATING THE USE OF TWO ELECTRONIC SWITCHES TO OBTAIN THREE TRACES THIS IS THE LIMIT FOR A GOOD DISPLAY ON A FIVE INCH SCREEN.

FIG. 36

37

Limiter Circuit Tests

The oscilloscope is an excellent test instrument for checking limiter circuits. Sometimes referred to as "clippers," limiters are used as "waveshapers" in many types of electronic equipment. A limiter stage can be an overdriven triode, a series diode, or parallel diodes. See Fig. 37. They are used to remove positive or negative peaks as a rectifier or to cut off the peak extremities when squaring a sine wave. Some limiter circuits are designed to eliminate noise pulses or level off the peaks of an FM carrier prior to detection. They also are used in TV receiver circuits to separate the sync pulses from the composite video signal and are designated "sync separators" or "sync clippers."

Procedure

Equipment Required:

Oscilloscope
Signal generator
Test circuits

The setup in Fig. 37A shows a parallel diode limiter circuit that clips off a portion of each peak of the applied cycle (see Oscillogram A). By careful adjustment the input sine wave can be displayed on a calibrated graticule and the amount of positive and negative peaks removed can be measured. Both peaks should be clipped equally. Although not intended as such, it offers a good test for matching a pair of diodes. Fig. 37B shows the use of two 20-volt Zener diodes back to back that conduct only during the extremities of the applied sine wave (see Oscillogram B).

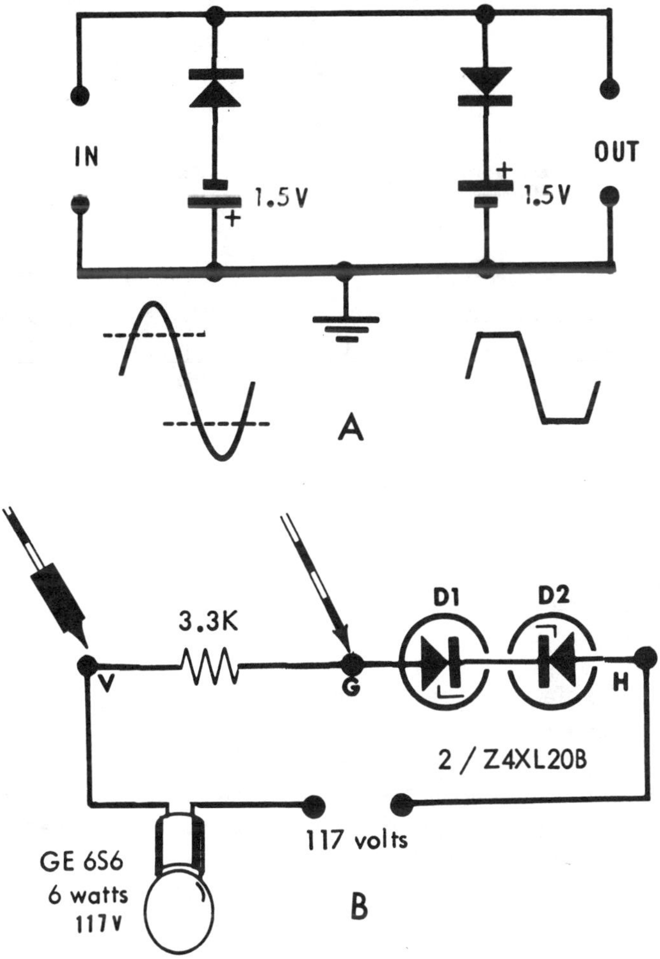
IN
OUT
1.5V
1.5V
A
3.3K
D1
D2
V
G
H
2 / Z4XL20B
117 volts
GE 6S6
6 watts
117V
B

FIG. 37

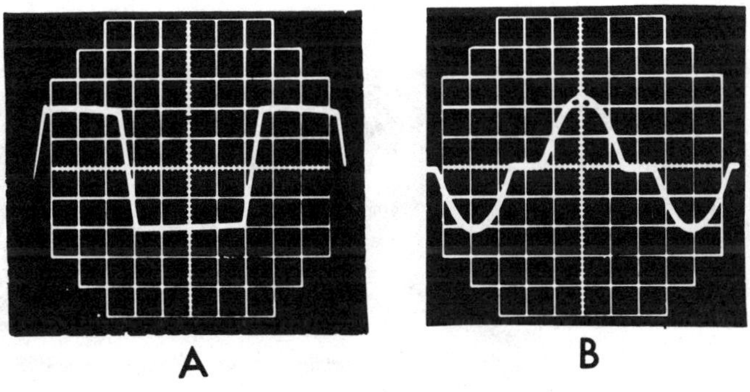
A
B

38

Checking Clamp Circuits

Clamping circuits are used to restore the DC component to an AC waveform after passing through a coupling capacitor. Two such clamping circuits are shown in Fig. 38.

Procedure

Equipment Required:

DC oscilloscope
Sine-wave generator
Test circuits

Connect the sine wave generator to the input of either a positive or negative clamping circuit and tune it to about 500 Hz. Adjust the sweep trace for zero reference, then connect the vertical input of the oscilloscope to the output of the clamping circuit. The DC average of an AC signal after it has passed through a coupling capacitor is zero, regardless of its amplitude. The average value of the same signal when passed through a clamping circuit will vary with the signal amplitude, generally referred to as a DC component that can be recovered when rectified. The oscillograms show both negative and positive clamps.

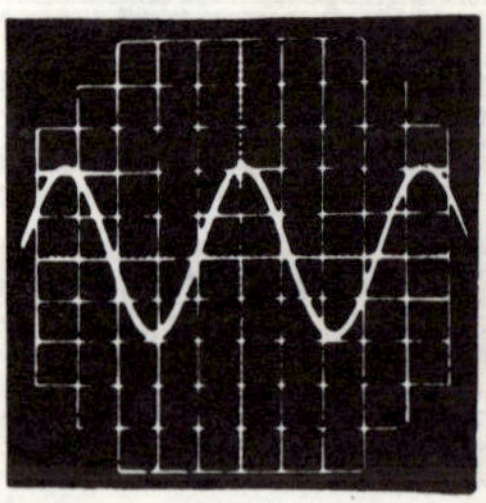

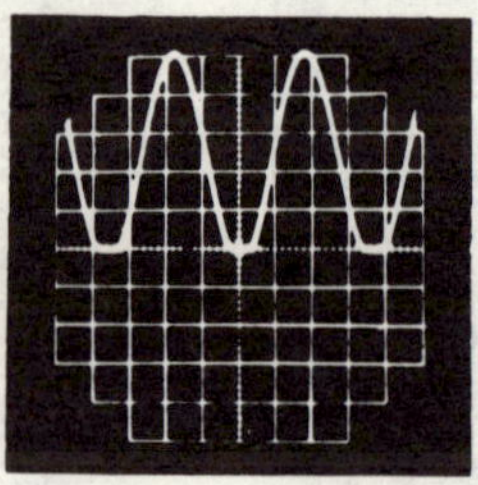

Positive clamp

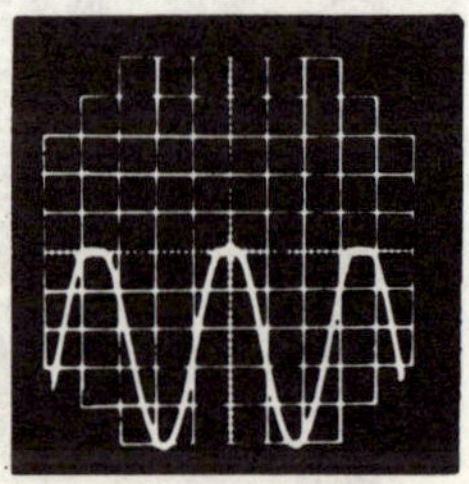

Negative clamp

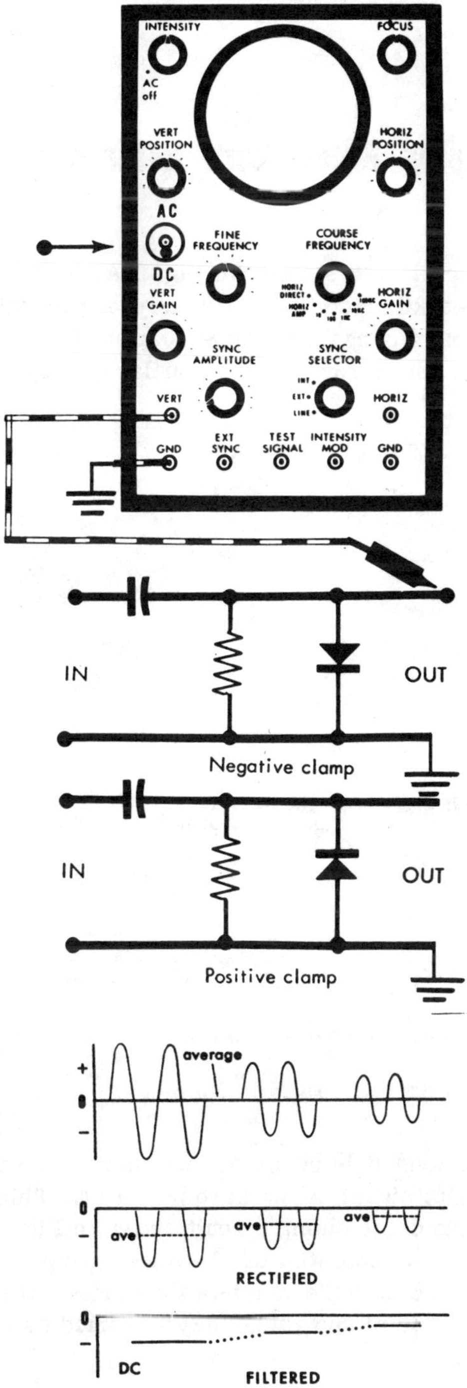
INTENSITY
FOCUS
AC
off
VERT
POSITION
HORIZ
POSITION
AC
DC
FINE
FREQUENCY
COURSE
FREQUENCY
VERT
GAIN
HORIZ
GAIN
SYNC
AMPLITUDE
SYNC
SELECTOR
VERT
HORIZ
GND
EXT
SYNC
TEST
SIGNAL
INTENSITY
MOD
GND
IN
OUT
Negative clamp
IN
OUT
Positive clamp
average
+
0
−
0
−
ave
ave
ave
RECTIFIED
0
−
DC
FILTERED

FIG. 38

39 Clamping Below Zero Reference

A triode clamping circuit is shown in Fig. 39A. The time constant of the series RC circuit is sufficiently long enough to hold the stage at cutoff during the negative portion of the input signal. In television this circuit is very popular as a sync clipper.

FIG. 39A

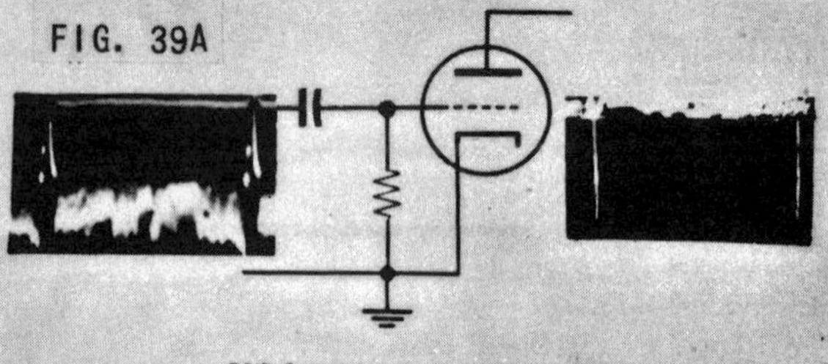

GRID CLAMPING CIRCUIT

FIG. 39B

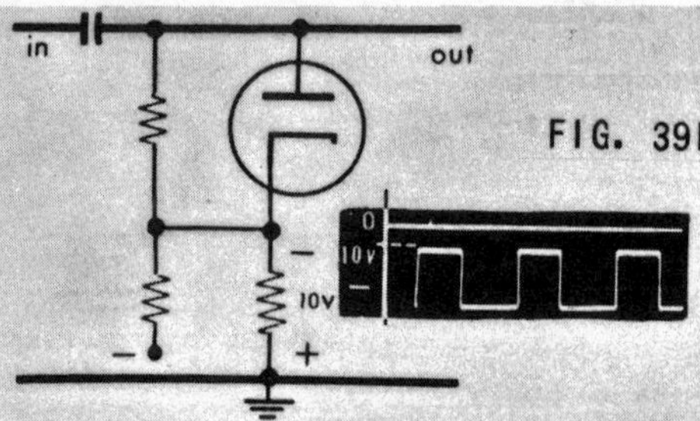

CLAMPING CIRCUIT ESTABLISHES A BELOW ZERO REFERENCE

In some applications it is necessary to clamp the upper extremity of an input signal below zero potential. This is accomplished by the diode clamp circuit shown in Fig. 39B, in this case using a DC potential of -10 volts. A typical application might be to clamp the start of the horizontal sweep of an oscilloscope. The clamp range may be varied by adjusting the DC potential.

40

Decreasing the Time Constant of Coupling and Its Effect

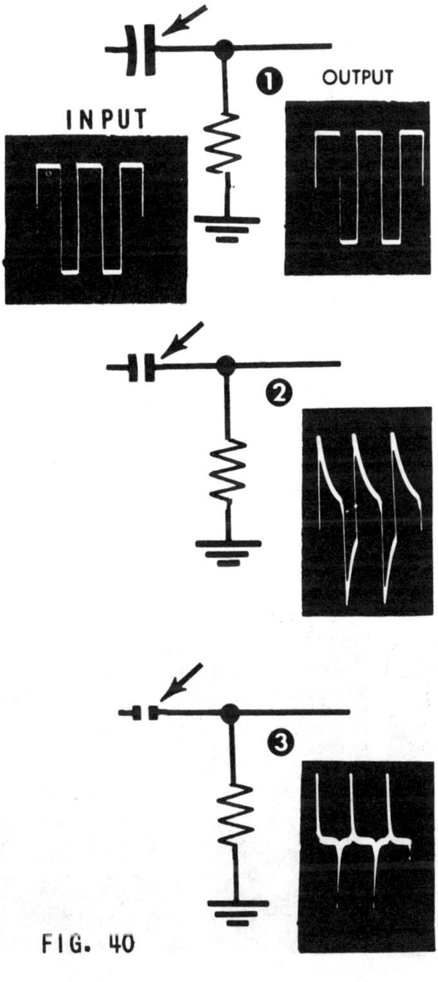

FIG. 40

Poor low-frequency response may be due to inadequate coupling capacity. This fact is illustrated in the following diagrams. DIAGRAM 1 shows a square wave applied to an RC coupling circuit. Notice the undistorted output. R1—500K; C—.25 mfd; square wave—500 Hz.

DIAGRAM 2 illustrates the effect of reducing the coupling capacitor from .25 mfd to .01 mfd. R1 remained unchanged.

DIAGRAM 3 illustrates the effect of further reducing C1 to .001 mfd. Notice that the leading and trailing edges are prominently displayed. These are the high-frequency components of the applied square wave. Thus a complete loss of low frequencies is obvious.

41

Detecting Hum Modulation

Hum modulation in AC-operated radio receivers is usually referred to a "tunable hum" and is due to a 60-Hz leakage in the RF or IF circuits. In practically all cases, tunable hum is caused by filament-to-cathode leakage. In some cases the hum is intermittent, making it difficult to track down since it could be in the audio section.

Procedure

Equipment Required:

Wideband oscilloscope
RF signal generator
Test circuit

Switch the signal generator to CW and tune to a frequency at the low end of the broadcast band. Connect a test lead to the output of the generator and place it near the receiver antenna; test for the presence of the signal by switching the generator

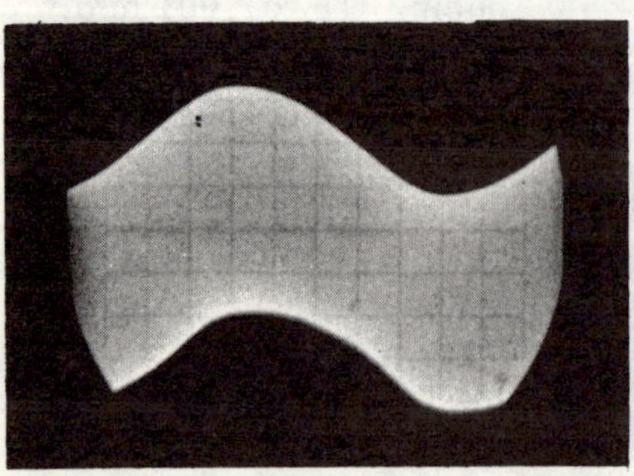

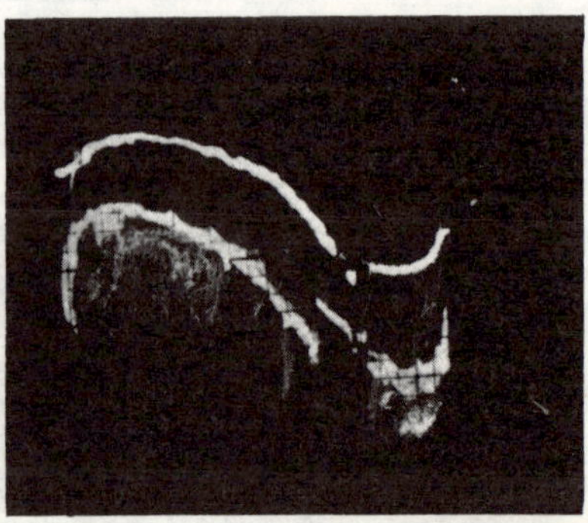

to modulation, then return to CW. Connect the oscilloscope vertical input probe to the plate of the last IF tube. If tunable hum is present it will show up as a wavy ribbon. The sweep frequency should be synchronized at line frequency (see Oscillogram A). A condition of hum modulation in a TV receiver is shown in Oscillogram B. This waveform was observed at the output of the video detector.

FIG. 41

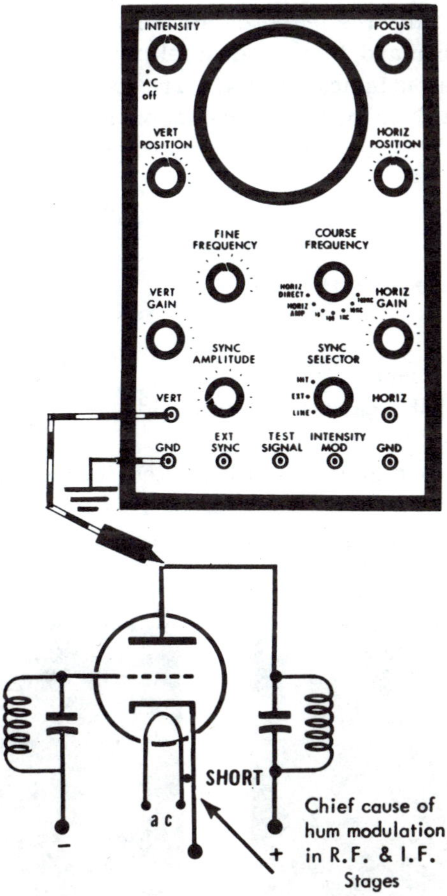

42

Hysteresis Loop: Single Coil

When iron is subjected to a varying magnetic field, the magnetism produced in the iron lags behind the applied magnetizing force. The greater the lag the greater the hysteresis loss appearing as heat. Tests show that soft iron and annealed silicon steel offer the least resistance to magnetic variations, while hard iron and tempered steel present the greatest opposition. This is indicated by the area between the two curves in Diagram A. The oscilloscope, being an ideal curve tracing instrument, can be used to display the B/H curve.

Procedure

Equipment Required:

Oscilloscope
Variac transformer
Resistor—150K; capacitor—0.1 mfd

Connect the test circuit as illustrated in Fig. 42 and adjust both horizontal and vertical gain controls to display the curve shown in the oscillogram. Vary the input voltage to observe the effect on the loop.

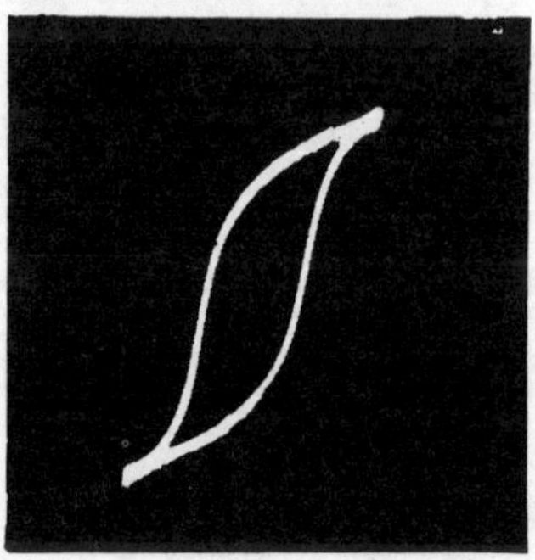

ANNEALED STEEL

SILICON STEEL

FIG. 42

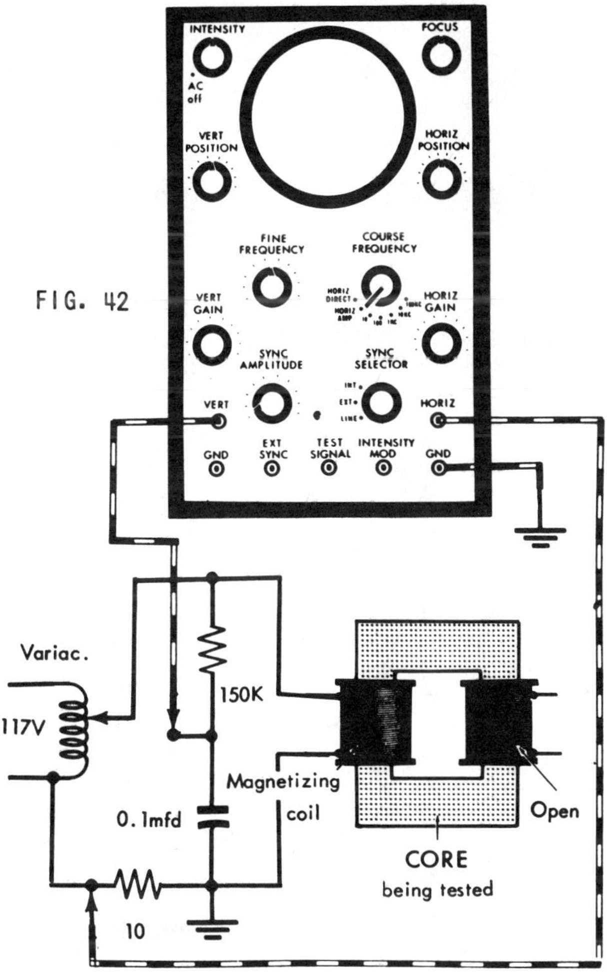

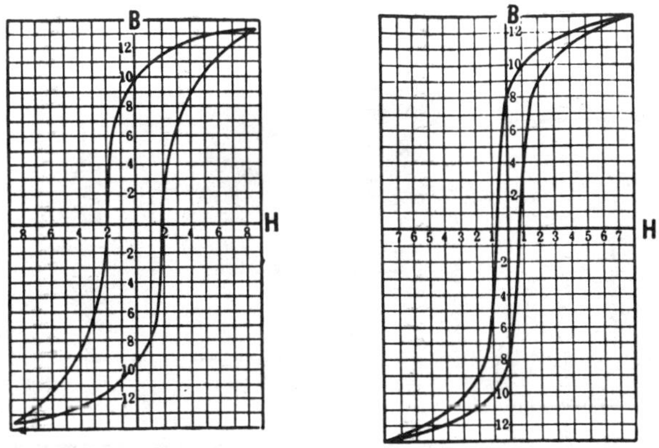

Poorly annealed steel

Well annealed steel

DIAGRAM A

43

Hysteresis Loop: Transformer

Fig. 43 shows connections required for checking an iron core in both primary and secondary windings. The hysteresis loop is the same as the primary connections.

FIG. 43

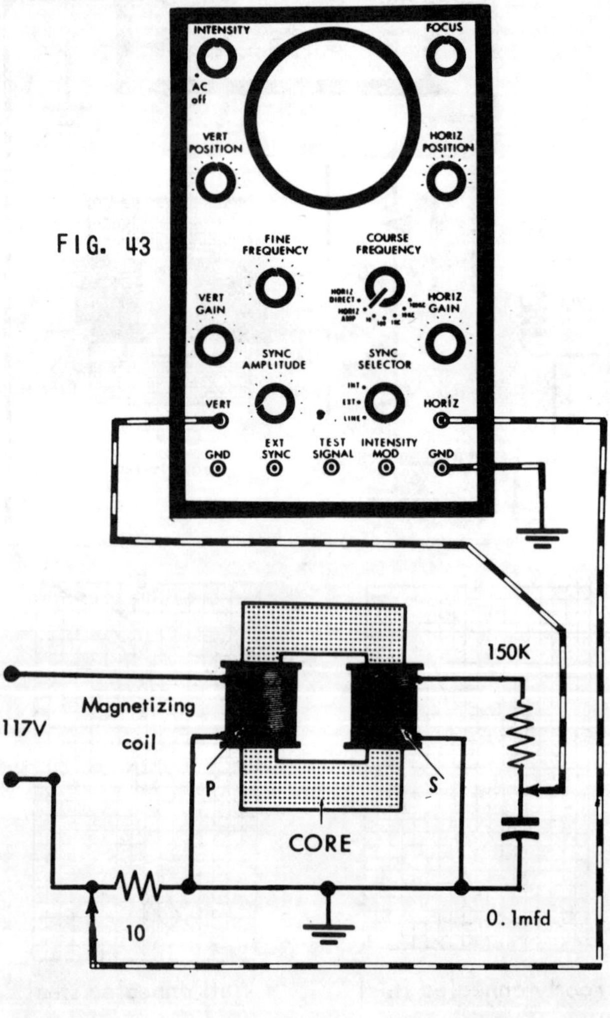

44

Step Counter Measurements

The 2-diode circuit shown in Fig. 44A delivers positive trigger pulses when a square wave is applied to the input. By reversing the diode connections the output becomes negative-going. Replacing R1 with C1 (see Fig. 44B), and by shunting the capacitor with Zener diode D1, the circuit becomes a step counter. After rectification by V2 each leading edge is integrated by the capacitor. Step integration continues until

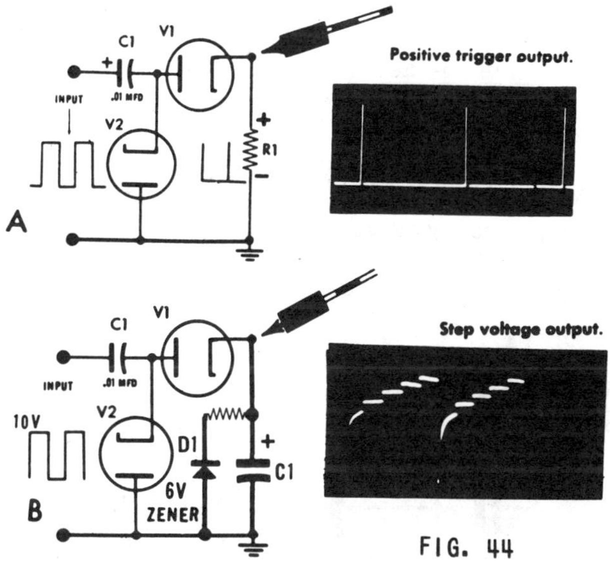

FIG. 44

C2 is charged sufficiently to cause the zener diode to conduct, which in turn discharges the capacitor and the cycle of events is repeated. The oscillogram shows four integrated triggers, called a staircase; at the end of each trigger there is a relatively large pulse—one pulse for four triggers, a division of 4 to 1. To obtain sharp triggers, connect the oscilloscope vertical input across the series resistor (about 3K).

45
46 Transistor Amplifier Types
47 and Characteristics

Three basic transistor configurations are shown in Figs. 45, 46, 47. To facilitate these tests, the input to each stage was adjusted to show three positive peaks and two negative peaks so that phase inversion could be identified more easily. The accompanying oscillograms are self-explanatory.

Fig. 45. Common-emitter amplifier: the collector waveform shows voltage gain and phase inversion; i.e., three negative peaks and two positive peaks.

Fig. 46. Common-base amplifier: The collector waveform shows voltage gain and no phase inversion; i.e., three positive peaks and two negative peaks.

Fig. 47. Common-collector amplifier: The collector waveform shows a voltage gain less than one and no phase inversion; i.e., three positive peaks and two negative peaks.

FIG. 45

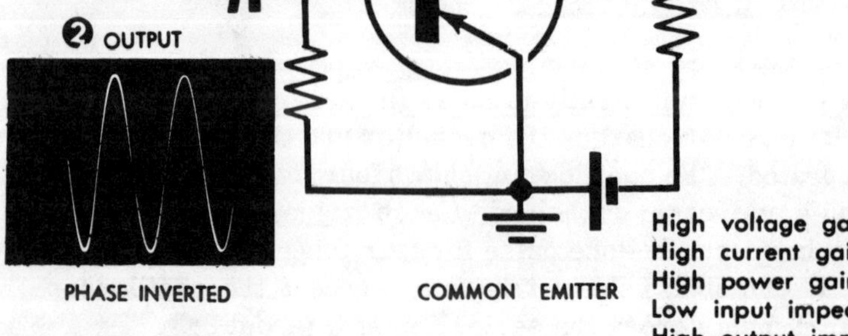

High voltage gain
High current gain
High power gain
Low input impedance
High output impedance

FIG. 46

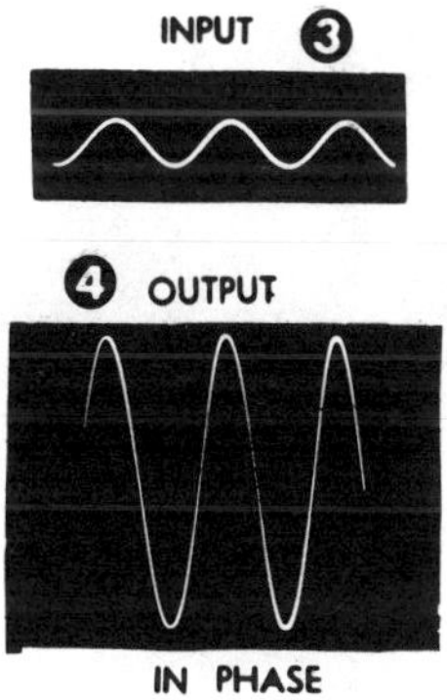

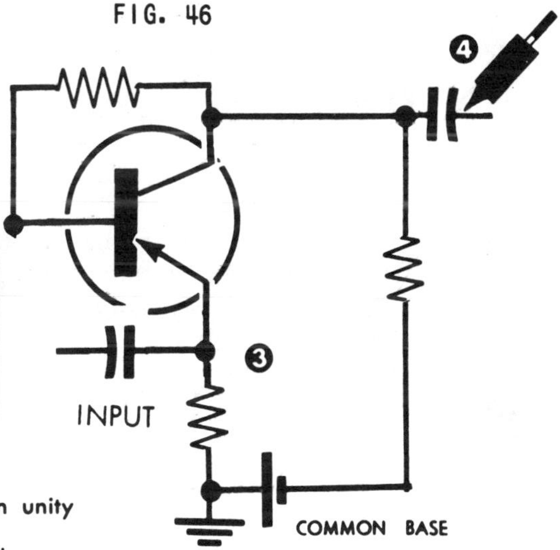

High voltage gain
Current gain less than unity
Medium power gain
Very low input impedance
Very high output impedance

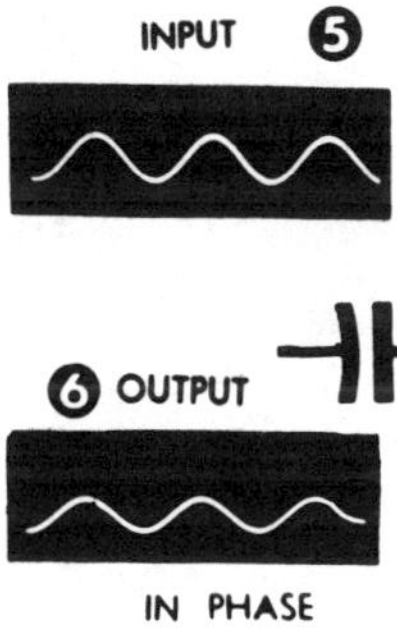

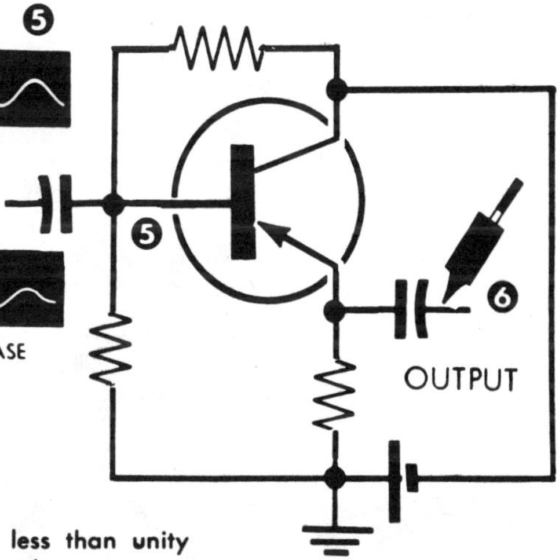

Voltage gain less than unity
High current gain
Low power gain
High input impedance
Low output impedance

FIG. 47

48

Checking Audio Amplifier Overload

Here is a simple test to determine whether or not an audio amplifier is overloaded. In this test a DC oscilloscope is necessary, and there must not be any phase difference between the vertical and horizontal amplifiers.

Procedure

Equipment Required:

DC oscilloscope
Audio signal generator

Connect the equipment as shown in Fig. 48 and tune the signal generator to 1 kHz at low volume. Turn up the gain control of both vertical and horizontal amplifiers and adjust each to provide a diagonal trace slanting at 45°, indicating equal traces. Slowly advance the signal generator output. If the trace begins to level off at one end it would indicate a form of overload.

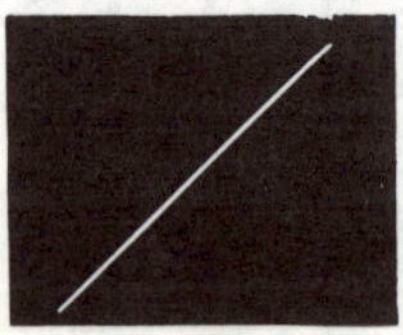

Normal

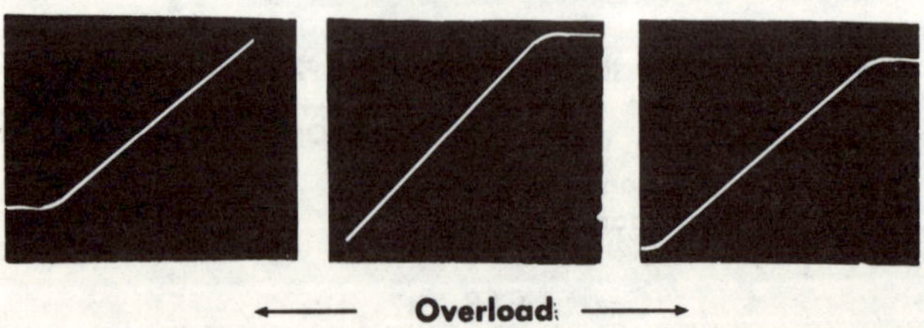

← Overload →

However, if both ends of the trace begin to level off simultaneously, it indicates the beginning of distortion due to overload. If one end of the trace begins to level off long before the other end, this indicates a bias defect or incorrect operating voltages (see the oscillograms).

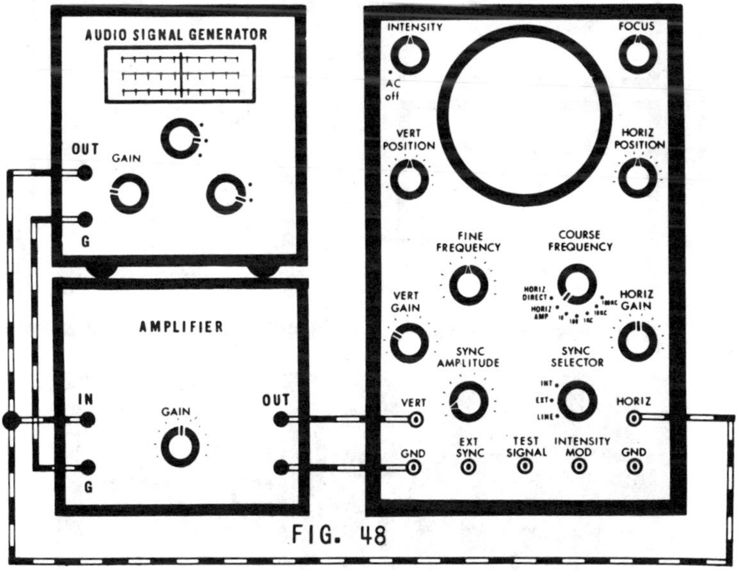

FIG. 48

---

49 Checking Audio Amplifier Response

Here is a simple but accurate method of determining the overall frequency response of an audio amplifier.

Procedure

Equipment Required:

Oscilloscope
Audio signal generator
Audio amplifier

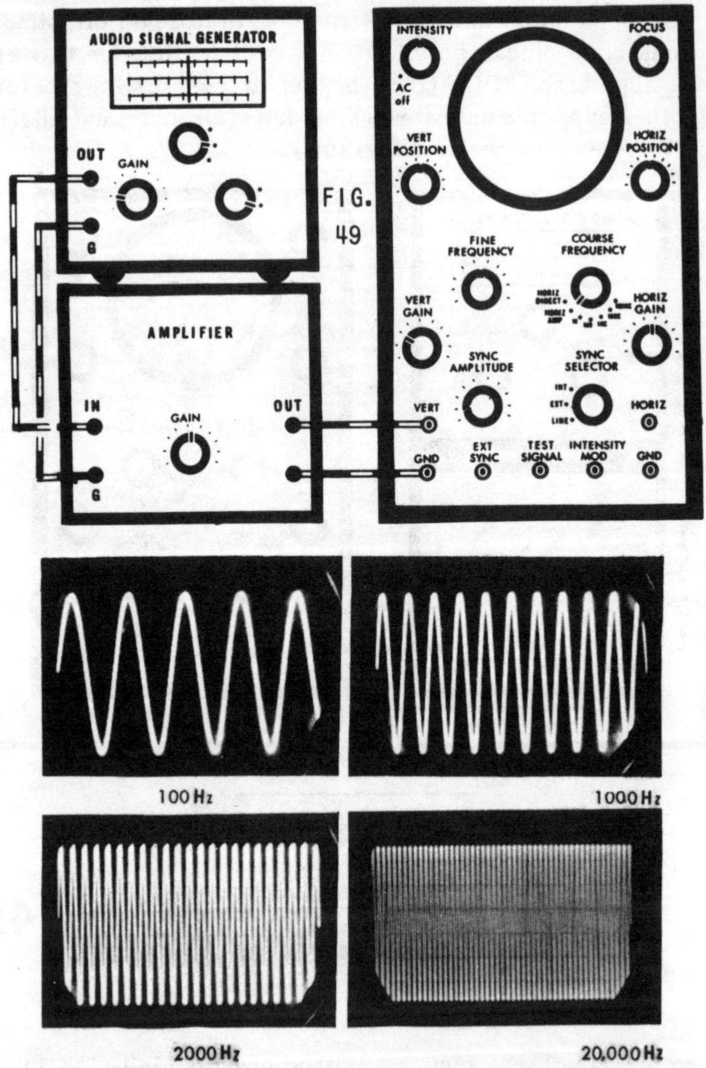

FIG. 49

Connect the signal generator output directly to the vertical input of the oscilloscope (1-kHz sweep rate) and check the audio response of both instruments before proceeding with the amplifier test. If the signal amplitude remains constant while tuning through 100 Hz to 20 kHz, check the amplifier as shown in Fig. 49 (see oscillograms). You'll notice a 10% reduction at 20 kHz, but this can be considered as a flat response over the entire audio range. With the test equipment set up, gain can be measured by using the test procedure in Fig. 34.

50

Checking Bandpass Amplifiers

The bandpass amplifier in a color receiver may be checked by measuring the pulses at the plate. Actually the killer pulse, which drives the amplifier into conduction, shock excites the 3.58-MHz tuned circuits and produces two damped waves for each pulse, an indication that the blanking amplifier, the killer stage and bandpass amplifiers are all in operating condition. No color signal is required for this check; in fact, this test should be conducted during a black-and-white transmission.

FIG. 50

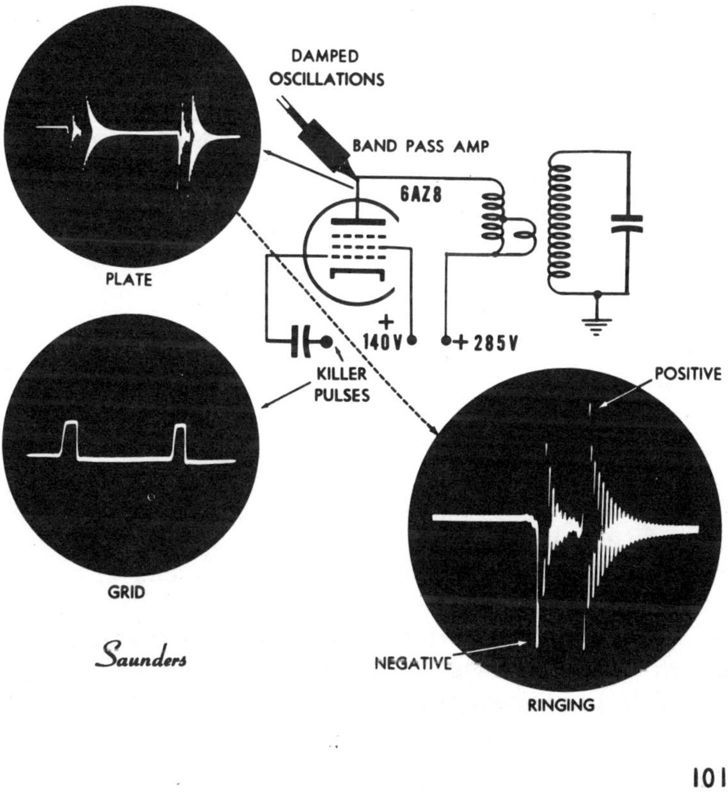

51

Color TV 3.58-MHz Oscillator Checks

In a color TV receiver the two subcarrier sidebands (R-Y and B-Y) appearing at the output of the bandpass amplifier are applied to the plate of their respective demodulators. From a locally generated 3.58-MHz crystal-controlled oscillator two subcarriers in quadrature (90°) are reinserted at each demodulator control grid. In the process the color sidebands and reinserted subcarrier are synchronized, resulting in demodulated chrominance signals.

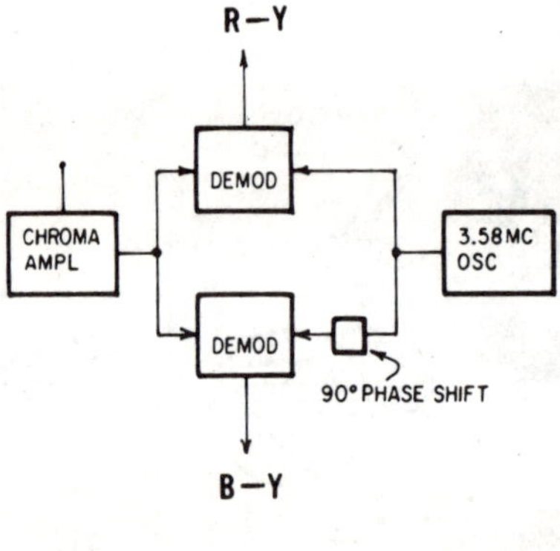

Procedure

Equipment Required:

Wideband oscilloscope
Low-capacity probe
Electronic switch (optional)

Set up the test equipment as shown in Fig. 51 and connect the electronic switch leads to the demoduators. Input A is connected to the control grid of the B - Y demodulator and input

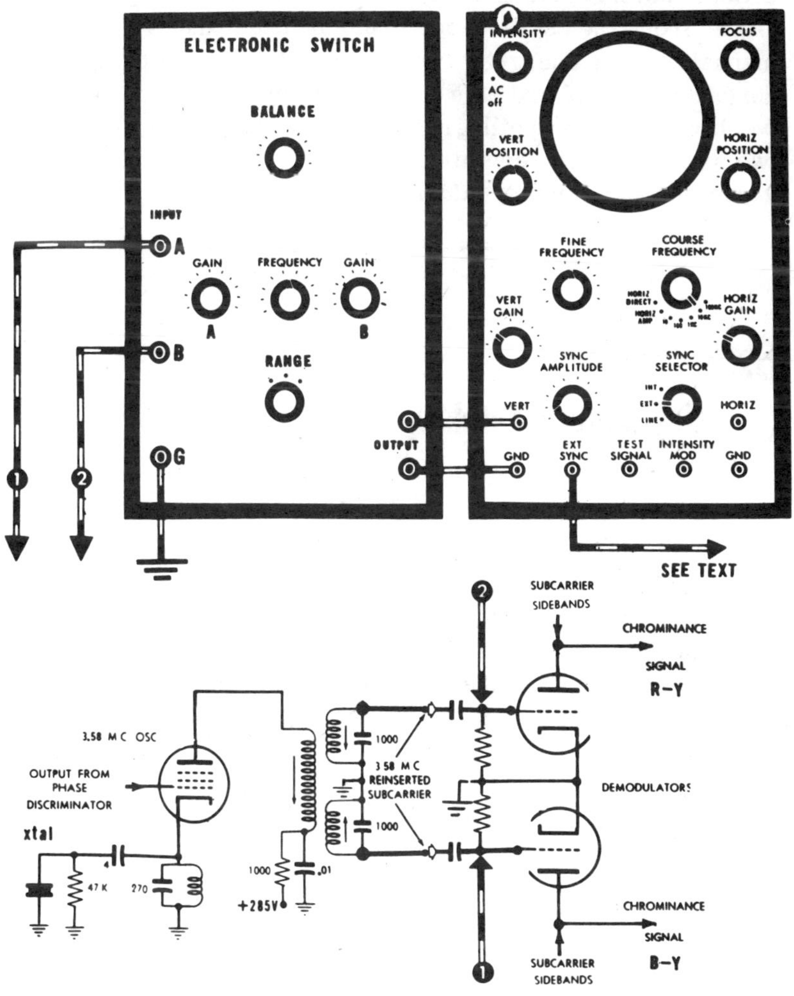
ELECTRONIC SWITCH
BALANCE
INPUT
A
GAIN
FREQUENCY
GAIN
A
B
B
RANGE
G
OUTPUT
INTENSITY
AC
off
FOCUS
VERT
POSITION
HORIZ
POSITION
FINE
FREQUENCY
COURSE
FREQUENCY
VERT
GAIN
HORIZ
GAIN
SYNC
AMPLITUDE
SYNC
SELECTOR
VERT
HORIZ
GND
EXT
SYNC
TEST
SIGNAL
INTENSITY
MOD
GND
SEE TEXT
1
2
SUBCARRIER
SIDEBANDS
CHROMINANCE
SIGNAL
R—Y
3.58 M C OSC
OUTPUT FROM
PHASE
DISCRIMINATOR
xtal
1000
3.58 M C
REINSERTED
SUBCARRIER
1000
DEMODULATORS
4
47 K
270
1000
.01
+285V
CHROMINANCE
SIGNAL
B—Y
SUBCARRIER
SIDEBANDS

FIG. 51

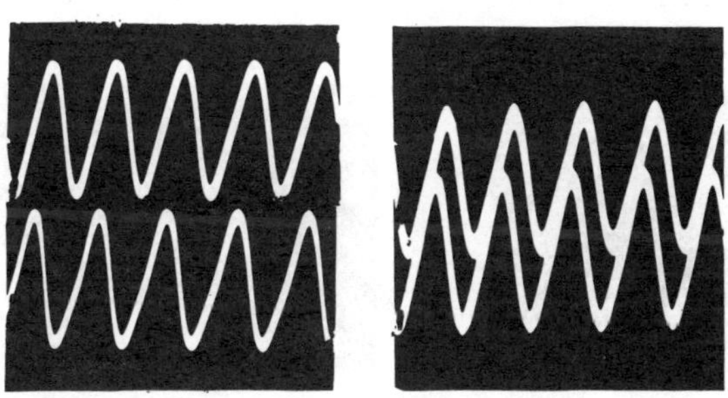

B to the control grid of the R-Y demodulator. A single probe direct from the oscilloscope could be used and transferred manually from one grid to another. Use the highest horizontal sweep rate and adjust downward until a few cycles can be observed. The oscillograms show the dual trace where the two signals can be seen in quadrature. The same test may be accomplished by using a dual trace oscilloscope.

52

Horizontal Blanking Pulse (Color TV)

The horizontal blanking pulse must not be confused with the horizontal retrace blanking pulse. The blanking pulse under discussion is derived from the flyback transformer and applied to the screen grid of the Y amplifier. This pulse is delayed and timed to blank out the burst signal interference that would appear on the left side of the raster as a vertical bar.

Procedure

Equipment Required:

Wideband oscilloscope

Check for the presence of the pulse at both the screen grid of the Y amplifier and at each CRT cathode. It should appear as a strong pulse, capable of cutting off the three beams. See the oscillograms.

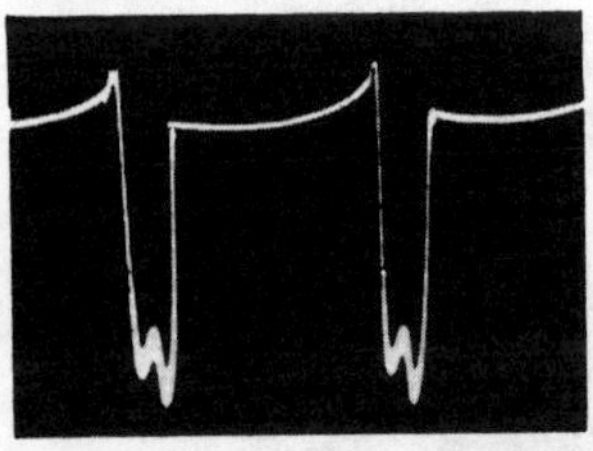

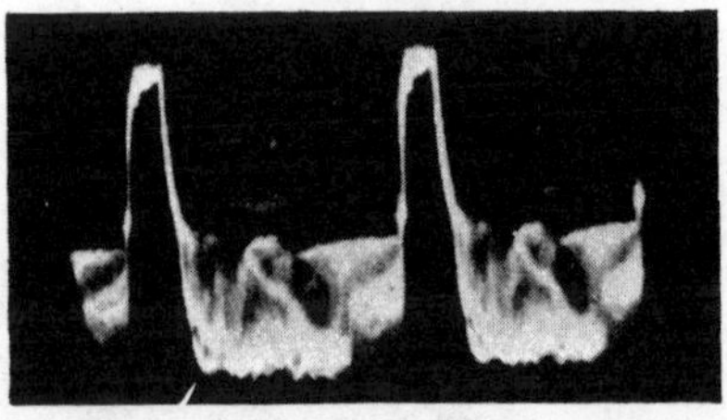

FIG. 52

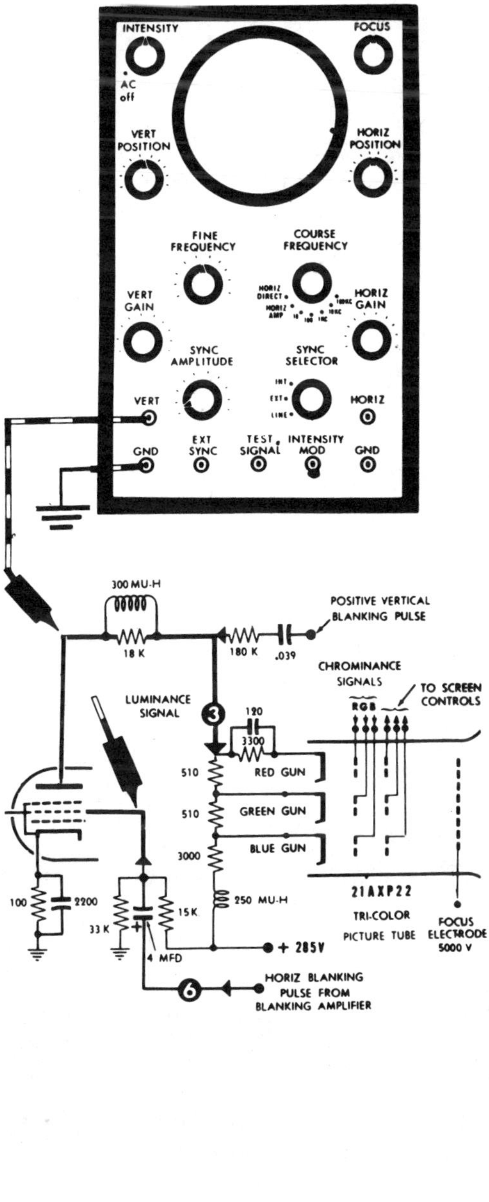
INTENSITY
AC
off
FOCUS
VERT
POSITION
HORIZ
POSITION
FINE
FREQUENCY
COURSE
FREQUENCY
VERT
GAIN
HORIZ
GAIN
SYNC
AMPLITUDE
SYNC
SELECTOR
VERT
HORIZ
GND
EXT
SYNC
TEST
SIGNAL
INTENSITY
MOD
GND
300 MU-H
18 K
180 K
.039
POSITIVE VERTICAL
BLANKING PULSE
CHROMINANCE
SIGNALS
RGB
TO SCREEN
CONTROLS
LUMINANCE
SIGNAL
3
120
3300
510
RED GUN
510
GREEN GUN
3000
BLUE GUN
250 MU-H
21AXP22
TRI-COLOR
PICTURE TUBE
FOCUS
ELECTRODE
5000 V
100
2200
33 K
15 K
4 MFD
+ 285V
6
HORIZ BLANKING
PULSE FROM
BLANKING AMPLIFIER

53

Checking R-Y, B-Y, and G-Y Signals

The color demodulator outputs drive three amplifiers—R-Y, B-Y, and G-Y respectively, and the amplified signals are then applied to their respective picture tube control grids.

Procedure

Equipment Required:

Oscilloscope
Color bar signal generator

Set the oscilloscope sweep rate to 7875 Hz. Connect the color generator to the antenna input and switch the receiver to an

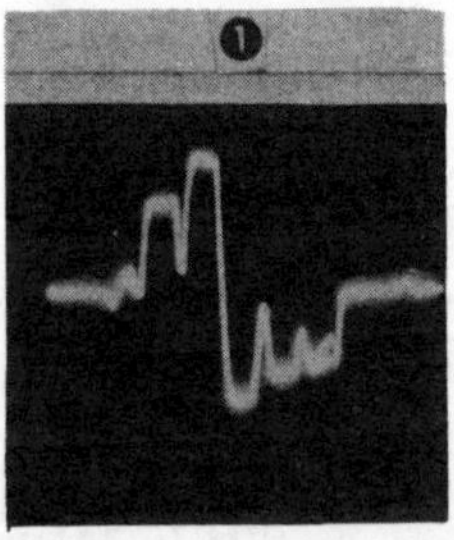

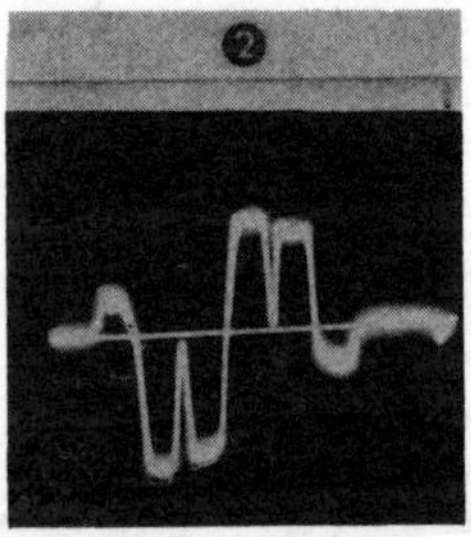

3

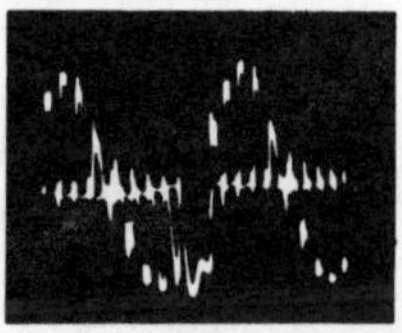

R-Y

4

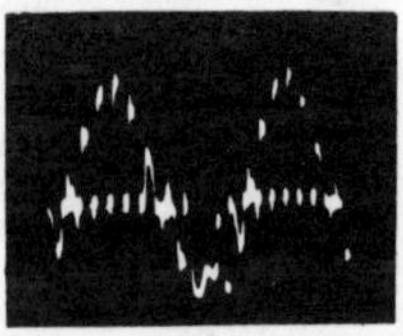

B-Y

5

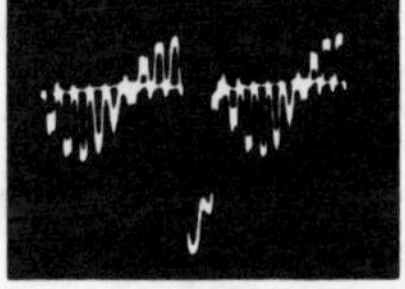

G-Y

unused channel or connect the generator R-Y, B-Y, and G-Y video output to the video input of the receiver, whichever is convenient. The color bar signals should appear at the control grid representing the signal to be observed. See Fig. 53. Oscillogram 1 is the G-Y signal and Oscillogram 2 is the R-Y signal, each obtained from a color generator in which the bars range from yellow to blue in order of their luminance. Another pattern using a generator with a different color sequence is shown in Oscillograms 3, 4, and 5.

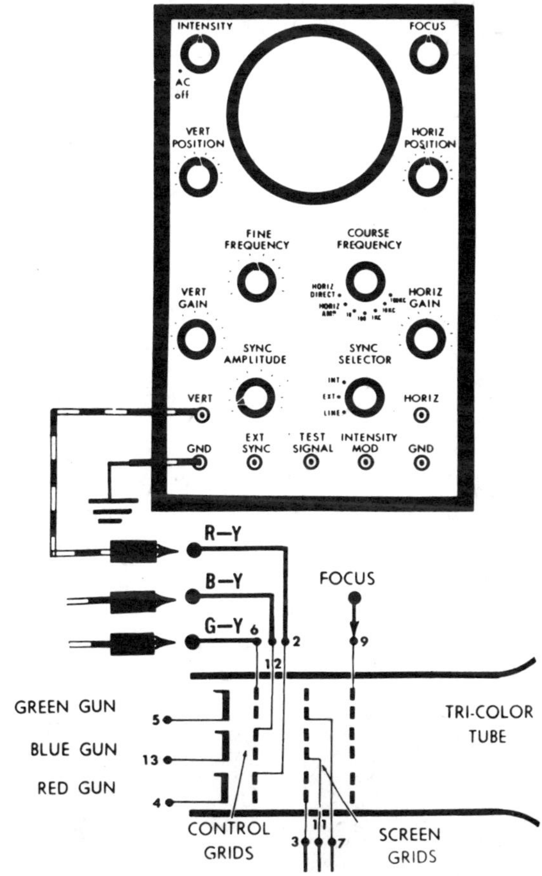

FIG. 53

54

DC Restorer

A composite TV signal carries sync, video, sound, and a DC component for the picture brightness level. However, when the composite signal is coupled to the picture tube by a capacitor it loses its DC component and the picture will have the same average brightness regardless the brightness of the transmitted scene. The DC component appears at the output of the video detector, varying with amplitude. Therefore, if a coupling capacitor is used between the video amplifier and the picture tube, a DC restorer must be incorporated at the input to the picture tube. See Fig. 54. The loss of the DC component also will have some effect on the sync pulses.

Procedure

Equipment Required:

DC oscilloscope

Check the oscilloscope for DC balance as suggested by the instrument manufacturer. Connect the oscilloscope vertical input probe to the picture tube video input terminal (usually the cathode) and tune in a TV station; set the oscilloscope sweep rate at 7875 Hz. If the restorer is operating correctly the sync tip level should remain stable despite the varying video amplitude. Use a picture with motion not a still one. See the oscillograms.

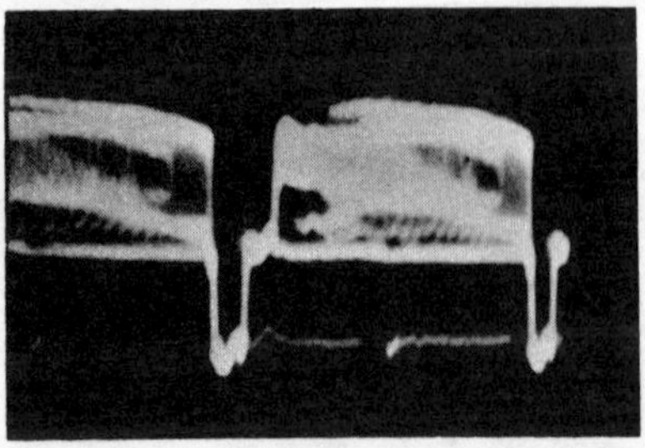

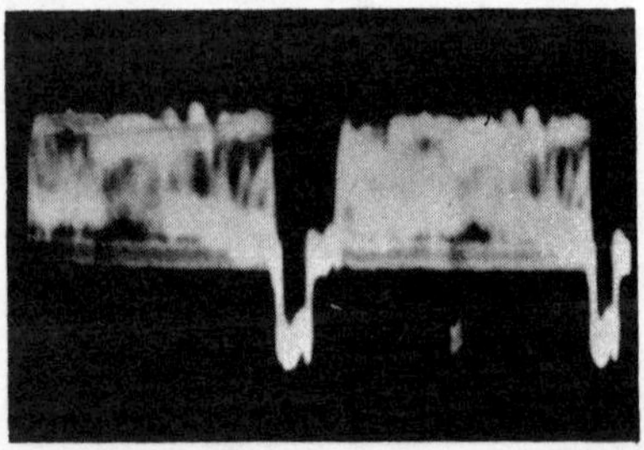

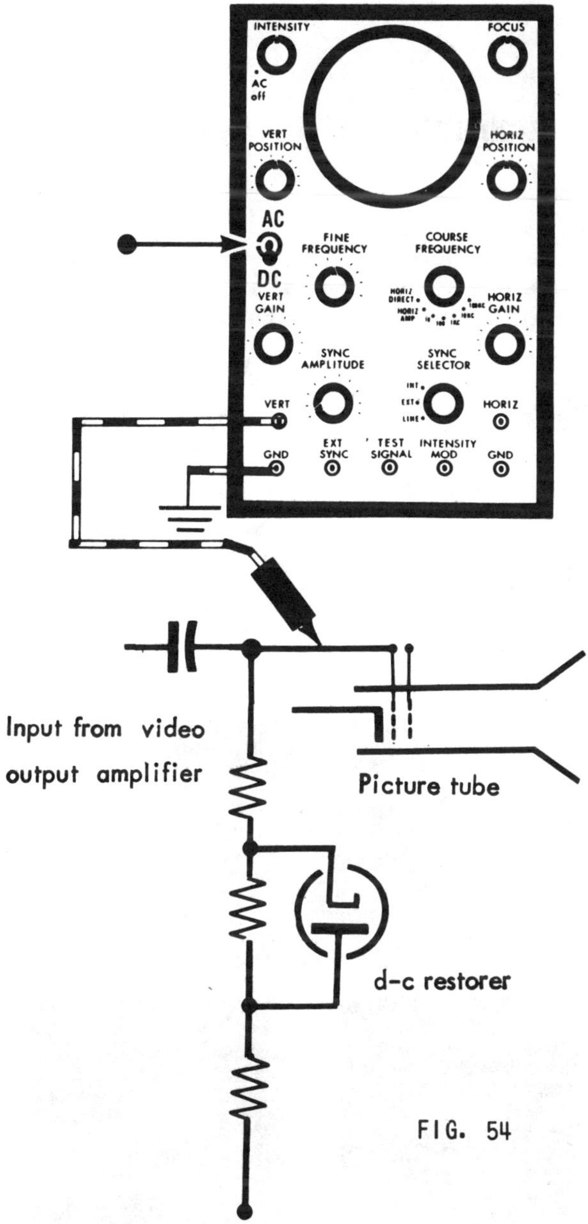
INTENSITY
FOCUS
AC
off
VERT
POSITION
HORIZ
POSITION
AC
DC
FINE
FREQUENCY
COURSE
FREQUENCY
VERT
GAIN
HORIZ
GAIN
SYNC
AMPLITUDE
SYNC
SELECTOR
VERT
HORIZ
GND
EXT
SYNC
TEST
SIGNAL
INTENSITY
MOD
GND
Input from video
output amplifier
Picture tube
d-c restorer

FIG. 54

55 Calibrating a DC Oscilloscope

DC oscilloscope calibration calls for a DC voltage source, just as an AC oscilloscope must be calibrated against a known AC voltage.

Procedure

Equipment Required:

DC oscilloscope
3K potentiometer
DC voltmeter

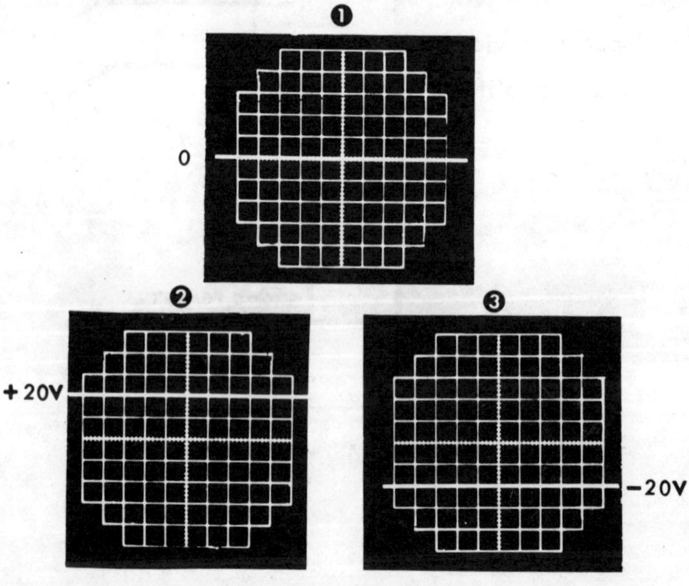

Calibration 10 Volts per square

Switch on the oscilloscope and allow an adequate warmup period. Adjust the horizontal positioning control until the sweep (1 kHz) is centered on the graticule. Connect the equipment as shown in Fig. 55 and adjust the potentiometer until the meter reads 20 volts. Then, adjust the vertical gain control until the sweep is two divisions above zero reference. When the vertical input leads are reversed the trace should be two divisions below zero reference. The oscilloscope is now calibrated to read 10 volts per division. For future reference label the vertical gain control with a marker.

FIG. 55

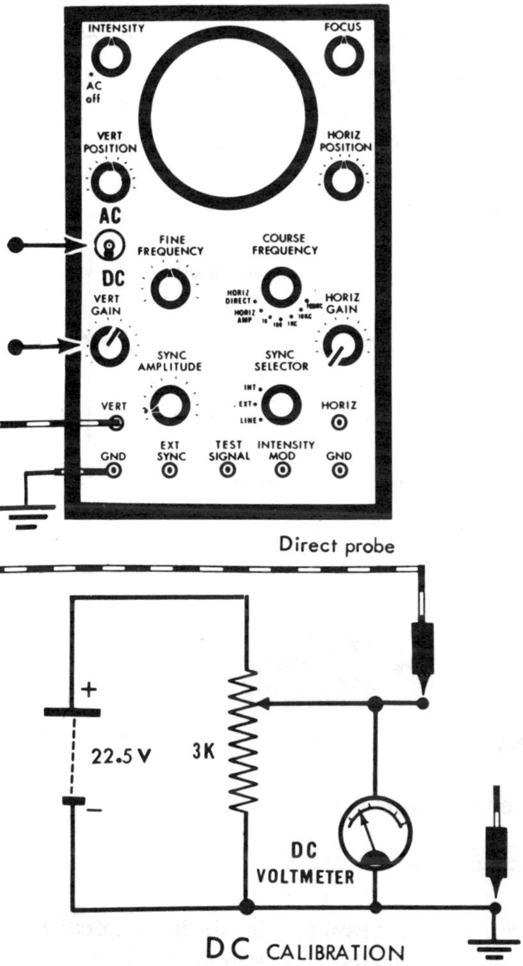

56 Electrostatic Deflection Tests

Practically all oscilloscopes use cathode - ray tubes designed for electrostatic deflection. The tube has two pairs of deflection plates mounted on the end of the electron gun. The potential applied to the horizontal pair must be a linear sawtooth wave, usually obtained from a multivibrator that generates an asymmetrical square wave of voltage which is applied to an output stage for amplification and waveshaping. When irregularities occur in the horizontal deflection system, a systematic check must be conducted.

electrostatic deflection

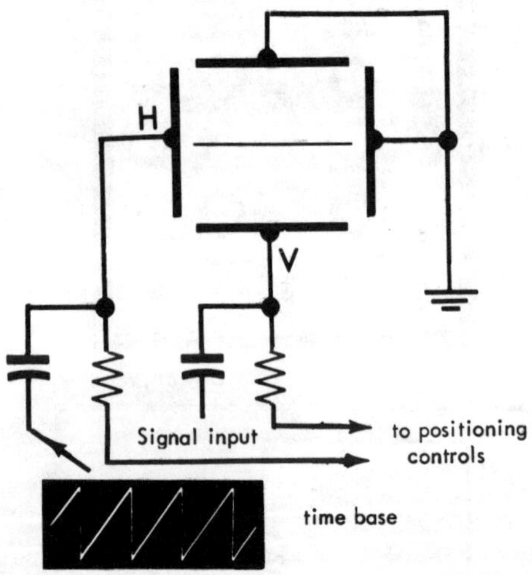

Procedure

Equipment Required:

Oscilloscope, direct probe

The schematic diagram in Fig. 56 is a typical sawtooth output amplifier; the input is an asymmetrical square wave which is amplified and shaped into a linear sawtooth sweep voltage by C4. The waveform at C4 is linear up to one third of the charging capacitor's time constant, which is approximately 20% of the applied voltage. Time constant is known as the period required for a capacitor to integrate 63% of the applied voltage. Hence, one third of the time constant would be about 20% of the applied wave.

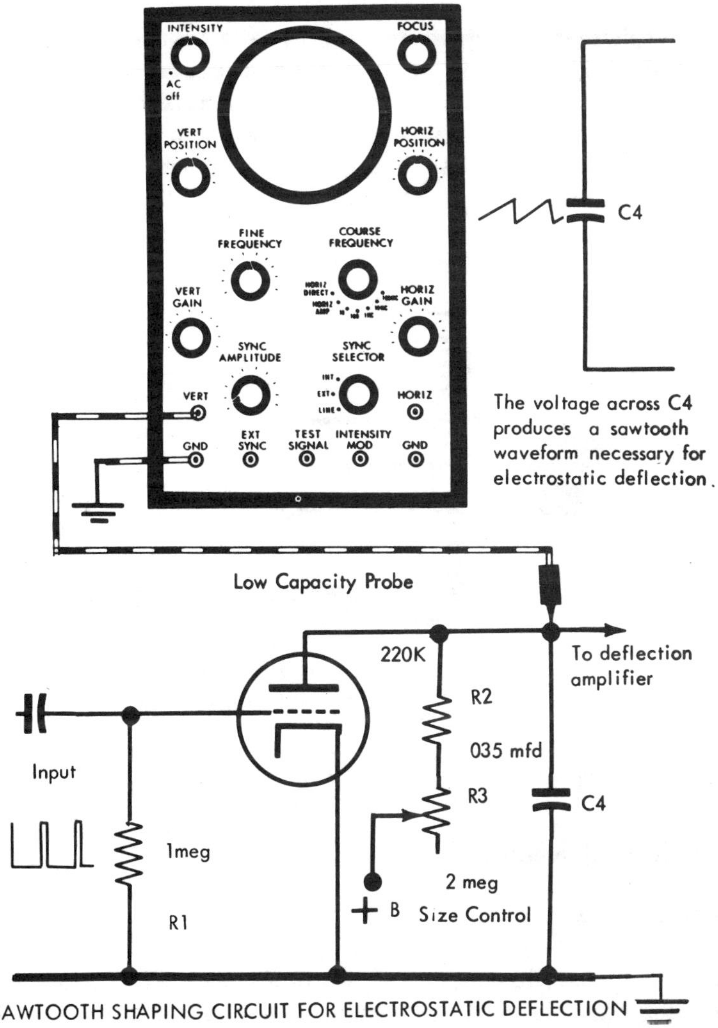

SAWTOOTH SHAPING CIRCUIT FOR ELECTROSTATIC DEFLECTION

FIG. 56

57

Electromagnetic Deflection Waveforms

The schematic diagram in Fig. 57 is a typical waveshaping circuit. The input is an asymmetrical square wave, and C4 and R4 provide the necessary time constant to produce the trapezoidal waveforms required for electromagnetic deflection. The peak-to-peak voltage and waveshape are important and should be checked when deflection troubles occur.

electromagnetic deflection

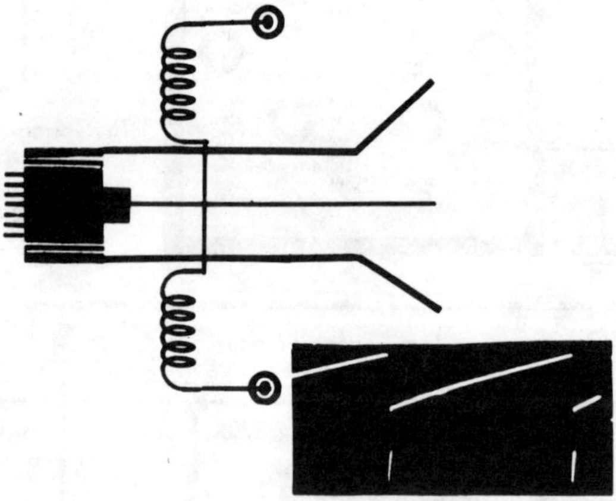

Procedure

Equipment Required:

Oscilloscope

Set up the oscilloscope as shown in Fig. 57. Adjust the positioning controls until the horizontal sweep is centered on the graticule. Connect the vertical input to the waveshaping amplifier plate and measure the peak-to-peak voltage and ratio of sawtooth to square wave. Voltage and waveshape should coincide with manufacturer's data. R3 controls amplifier gain, hence the amplitude of the waveform.

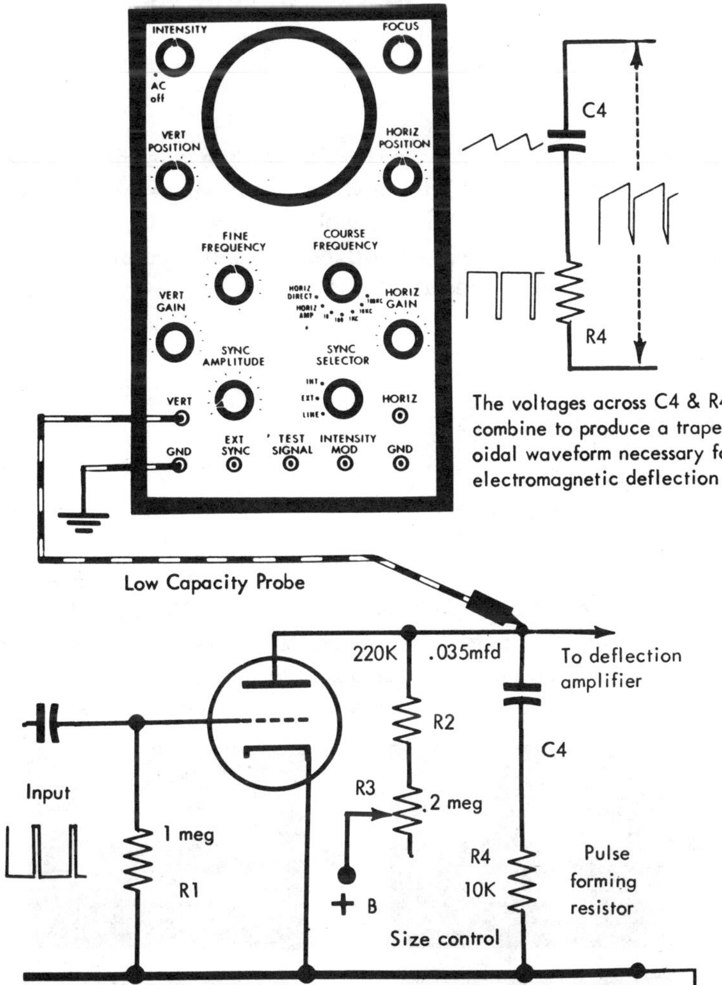

PULSE SHAPING CIRCUIT FOR ELECTROMAGNETIC DEFLECTION

FIG. 57

58

Vertical Oscillator Waveforms

The television receiver vertical deflection system consists of a synchronized oscillator and an output stage which drives the deflection coils. When trouble in this section exists, a check of the waveforms should be made and compared with the manufacturer's specifications.

Procedure

Equipment Required:

Oscilloscope
10-to-1 attenuator probe

Fig. 58 shows a typical vertical deflection system. The waveform at the oscillator grid is shown in Oscillogram 1. The positive peaks indicate short-duration conduction; also notice the vertical sync pulse on the positive peaks. At position 2, Oscillogram 2 shows the charge across the coupling capacitor

❶

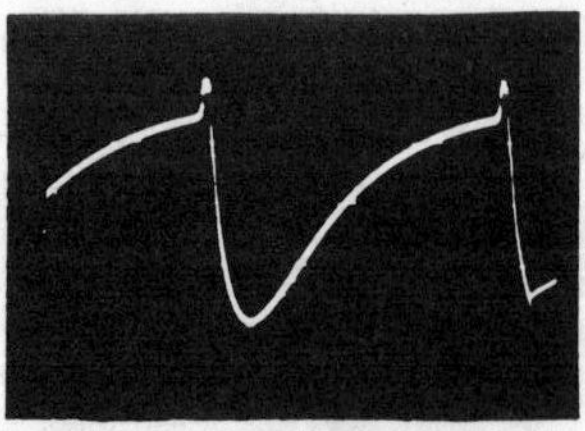

❸

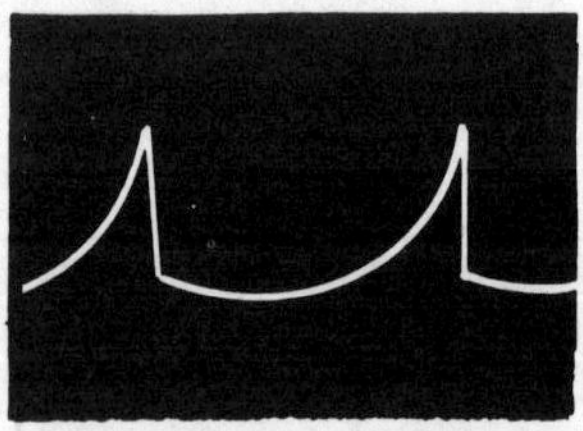

❷

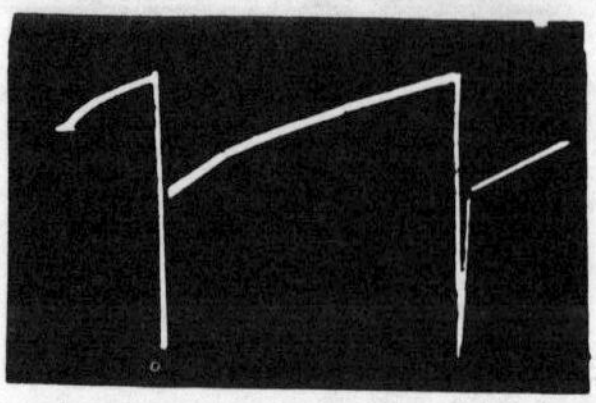

❹

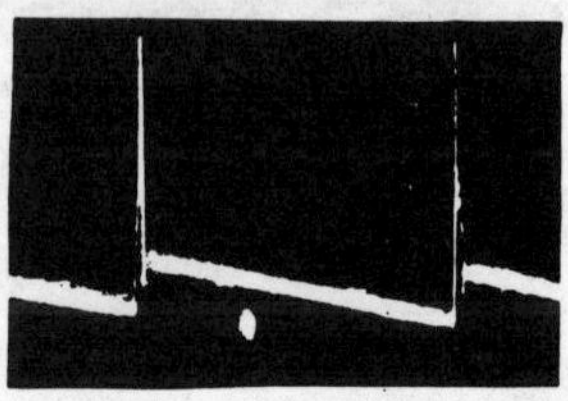

during oscillator cutoff and rapid discharge during conduction. The cathode waveform (position 3 and Oscillogram 3) shows a curvature which varies with linearity control adjustment. The waveform at the plate of the output stage (position 4) appears in Oscillogram 4. Notice that the trapezoidal waveform is reversed to that of the input, indicating current conduction through the deflection coils. The thick sawtooth portion is due to induction from the horizontal coils. This is normal and does not effect vertical deflection.

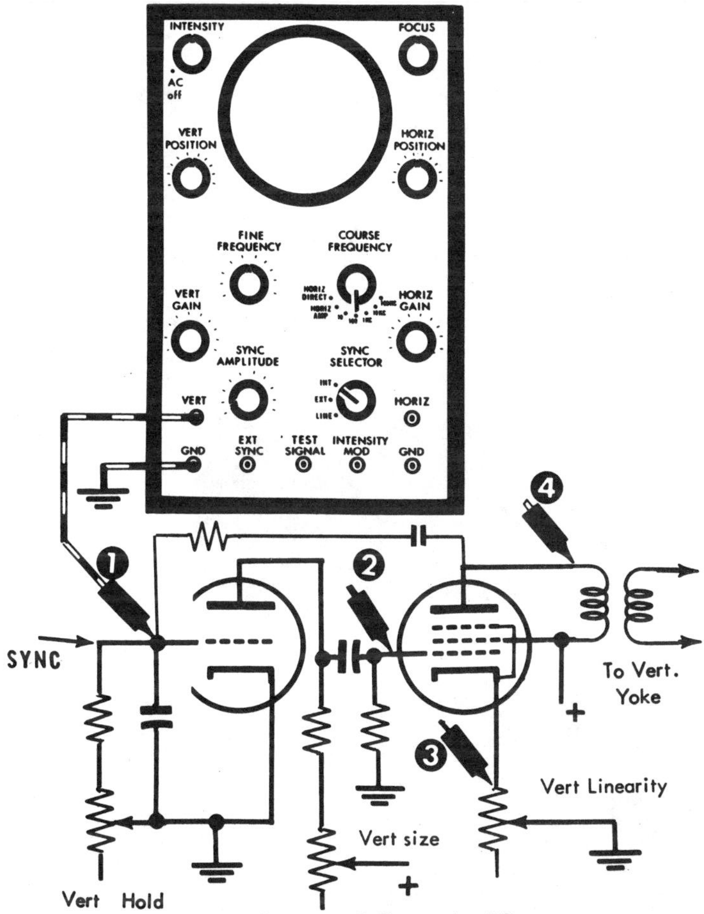

Vertical Oscillator & Output Amplifier

FIG. 58

59

Vertical Deflection Coil

To develop a sawtooth of current through a coil, the applied voltage must be a sawtooth/square-wave combination. If the coil were a pure inductance, a square-wave voltage would develop a sawtooth current. It is an established fact: to develop a sawtooth current waveform through a resistor a sawtooth waveform of voltage is necessary, but since a coil contains both inductance and resistance, the combination of both sawtooth and square wave is necessary. This waveform is often referred to as a "trapezoidal" waveform. The ratio of these two waveforms would depend upon the ratio of reactance to resistance. However, most waveforms tested have been found to be more square than sawtooth (see the oscillogram).

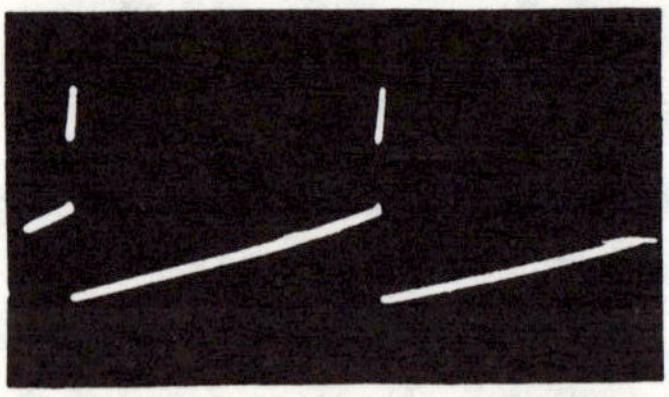

Procedure

Equipment Required:

Oscilloscope
.25 mfd capacitor

Connect the oscilloscope as shown in Fig. 59 and adjust the sweep rate to 60 Hz to obtain the trapezoidal waveform shown in the oscillogram. Remember: In these tests the oscilloscope housing is <u>350 volts</u> <u>above</u> ground!

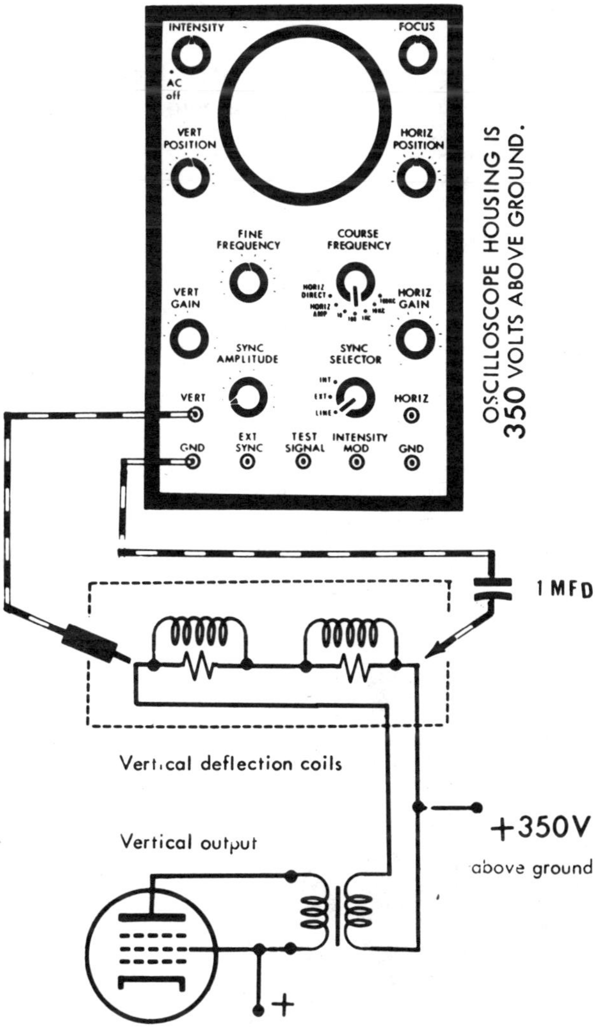
INTENSITY
FOCUS
AC
off
VERT
POSITION
HORIZ
POSITION
FINE
FREQUENCY
COURSE
FREQUENCY
VERT
GAIN
HORIZ
GAIN
SYNC
AMPLITUDE
SYNC
SELECTOR
INT
EXT
LINE
VERT
HORIZ
GND
EXT
SYNC
TEST
SIGNAL
INTENSITY
MOD
GND
OSCILLOSCOPE HOUSING IS
350 VOLTS ABOVE GROUND.
1 MFD
Vertical deflection coils
+350V
above ground
Vertical output
+

FIG. 59

60

Horizontal/Vertical Convergence Current Waveforms

The following tests indicate the presence or absence of the correct convergence waveforms in a color TV receiver.

Procedure

Equipment Required:

Oscilloscope

Insert a 10 - ohm resistor in series with the convergence coil lead and connect the oscilloscope across it (see diagram). Adjust the oscilloscope sweep rate to 30 Hz. The coil current will be in phase with the voltage developed across the resistor; hence, the waveforms will represent current flowing through the coil. Notice that the waveform contains both the vertical and horizontal convergence waveforms. (Oscillograms 2 and 4 were observed when receiver was in convergence.)

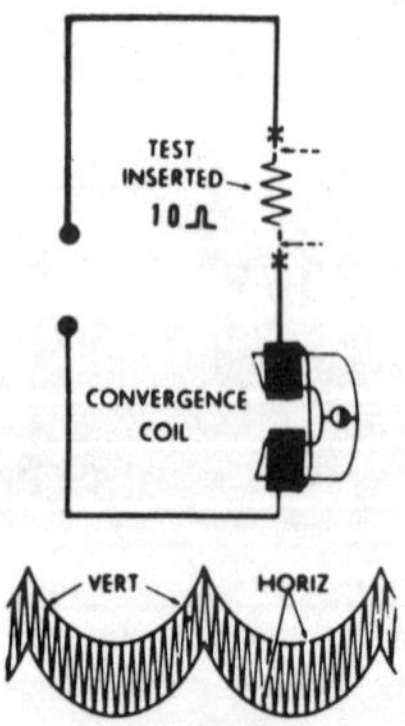

VERT & HORIZ PARABOLIC WAVESHAPES

FIG. 60

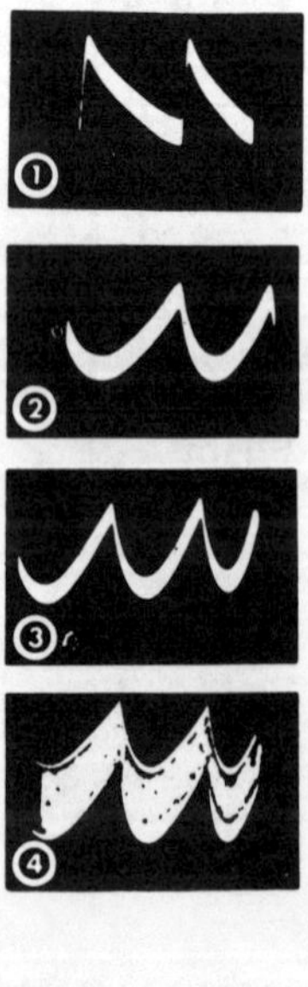

1. External vertical tilt

2. Correct vertical waveform

3. Reduced horizontal amplitude

4. Increased horizontal amplitude

61 Horizontal Deflection Current Waveforms

The current flowing in horizontal deflection coils is a linear sawtooth waveform. Checking the current waveform is equivalent to checking the whole deflection system, since any defect will ultimately show up in the waveform.

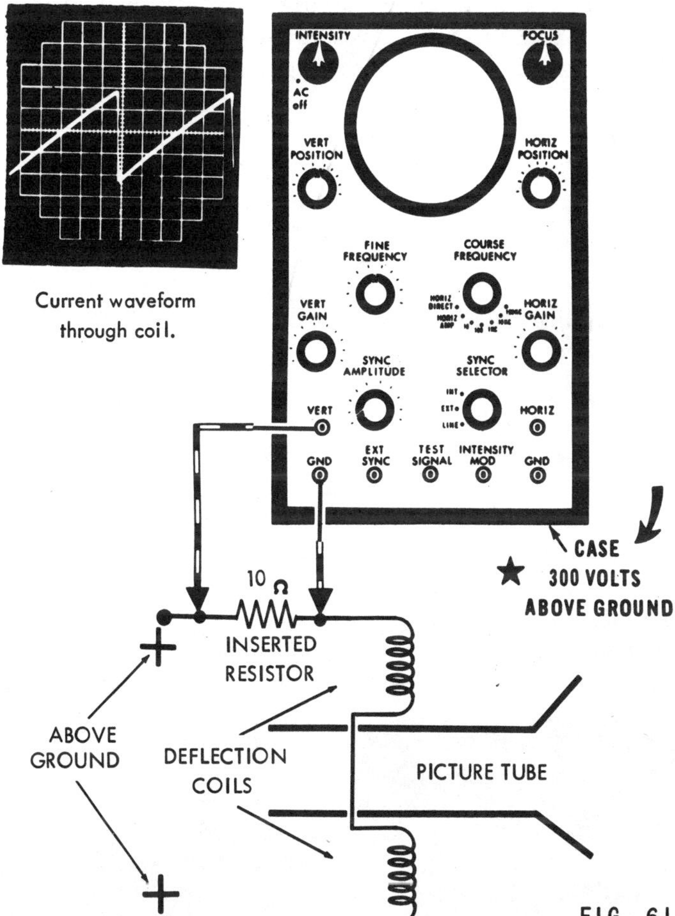

Current waveform through coil.

FIG. 61

Equipment Required:

Oscilloscope
10-ohm resistor

Insert a 10-ohm resistor in series with the deflection coil lead as shown in Fig. 61. Connect the oscilloscope vertical input across the resistor and adjust the sweep rate to 7875 Hz. Turn on the power and observe the waveform (see oscillogram). Precaution: When conducting this test it must be remembered that the oscilloscope housing is about 300 volts above ground. Take all necessary care to prevent bodily contact between the oscilloscope and receiver chassis. Inexperienced personnel should not conduct this test. Keep the power off while setting up the test.

---

62 Video Amplifier: Split Video Signal

As indicated in Fig. 62, the video amplifier distributes a number of composite signal components to various parts of the set. Notice that two samples of the video signal are distributed to the noise inverter stage: one is taken off the plate, which is inverted, and the other from the cathode which is not inverted. When servicing this section for any particular malfunction, it is a good practice to check the noise inverter control, since this is a threshold adjustment (see oscillograms).

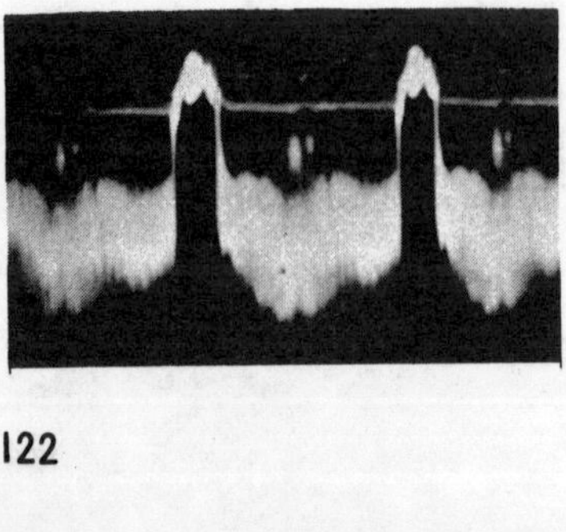

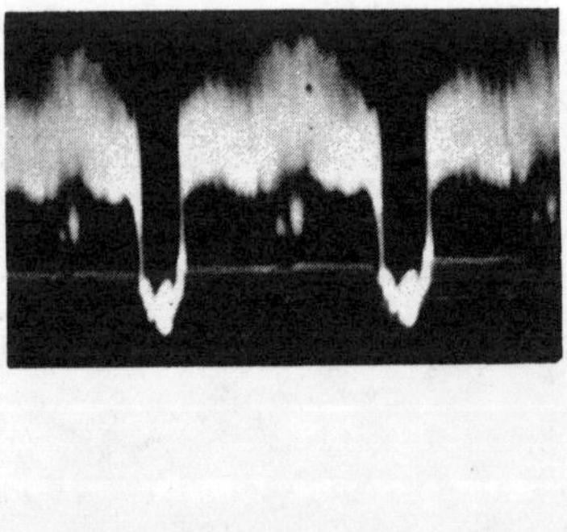

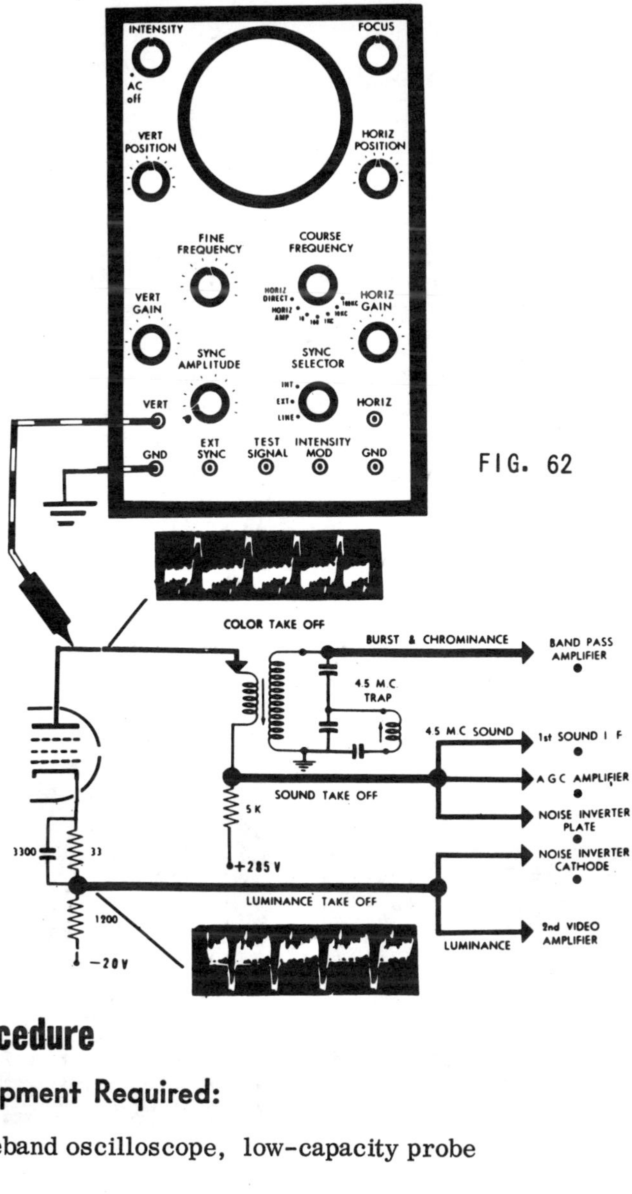

FIG. 62

Procedure

Equipment Required:

Wideband oscilloscope, low-capacity probe

Tune in a local TV station and connect the vertical input to the plate of the video amplifier and adjust the oscilloscope to display the composite signal (sweep rate 60 Hz and 7875 Hz). Examine both vertical and horizontal sync pulses and check all distribution points. Repeat this test by connecting probe to the cathode of the video stage.

63

Checking Video Amplifier Gain

Without a test signal the following method may be used as a quick means of ascertaining video amplifier gain. Tune in a signal from a local TV station and compare the height of the sync pulses at the input of the amplifier to those at the output.

Procedure

Equipment Required:

DC oscilloscope
Low-capacity probe

❶

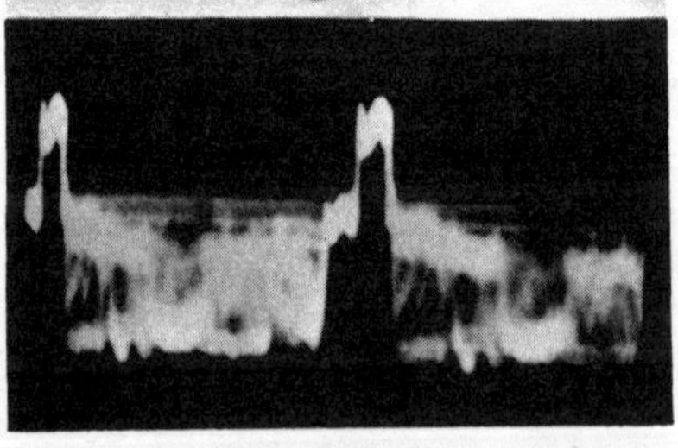

❷

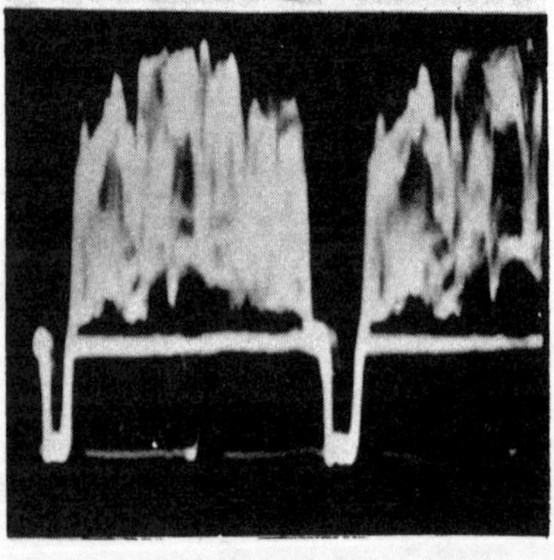

Set up the test as illustrated in Fig. 63 and connect the probe to the control grid of the video amplifier. Adjust the input attenuator to X1 and set the horizontal sweep to the 10-kHz range. Tune in a local channel and adjust the fine frequency control until two horizontal pulses appear (sweep rate 7875). Ignore video program material and measure the height of the sync pulse. Turn the vertical input attenuator to X100 and connect probe to the video amplifier plate. If the height of the output pulse is equal to the input sync pulse, then the gain is 100. In the accompanying oscillograms a little more than 100-to-1 gain was experienced.

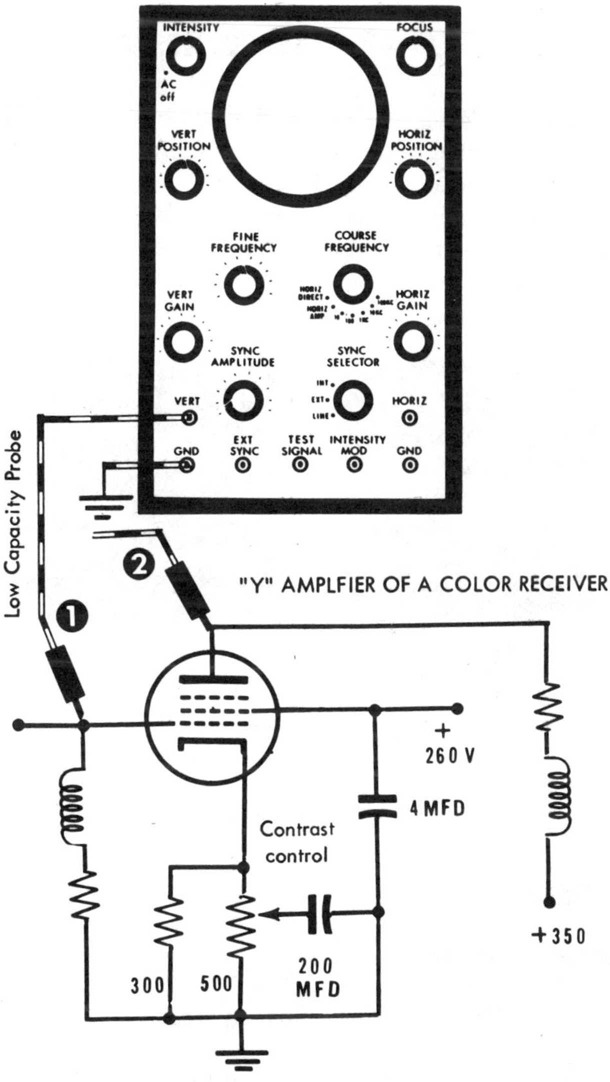

FIG. 63

64

Video Amplifier: Poor Low-Frequency Response

The video amplifier is usually the distribution point of all signal components; therefore it must be stable and have a broad frequency response. The screen grid bypass capacitor, usually 4 mfds, should be suspect when low frequencies fall off. Case histories have shown that the low-frequency response is impaired when this capacitor is defective, causing uncontrollable picture roll or intermittent loss of vertical sync.

❶ 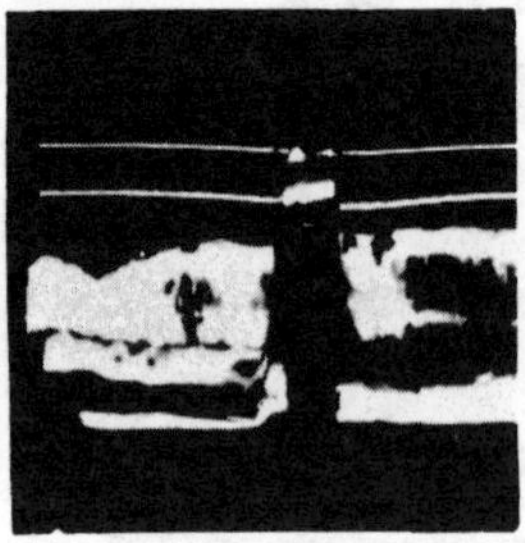

 ❷

1. Normal response
2. Poor low-frequency response Notice that the horizontal sync pulse is well above the 60-Hz vertical sync pulse.

Procedure

Equipment Required:

Wideband oscilloscope, low-capacity probe

Tune in a local TV station. Connect the vertical input to the plate of video amplifier and adjust the sweep rate to 60 Hz.

The vertical sync pulse should appear normal and equal with horizontal sync level. If some discrepancy is observed, connect the oscilloscope probe to the screen grid. If the video signal appears at this point, it would indicate a defective filter capacitor. Even if a very small amount of the video signal appears, it has been known to cause intermittent vertical roll.

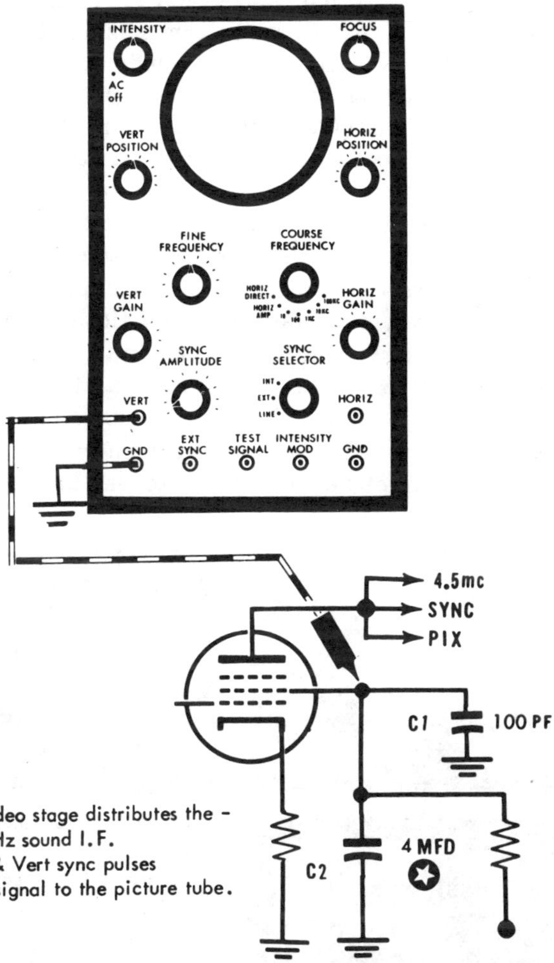

This video stage distributes the -
4.5 MHz sound I.F.
Horiz & Vert sync pulses
Video signal to the picture tube.

FIG. 64

65

Noise Inverter Circuits

Positive noise pulses that extend above the sync tips in the composite TV signal can be troublesome if permitted to enter the sync circuits. Noise spikes have high-frequency characteristics and can cause random triggering of the horizontal sweep oscillator. In the receiver under discussion a noise inverter stage is used at the input of the first sync amplifier and AGC amplifier. (See Fig. 65.) A negative-going composite signal is applied to the cathode of the noise inverter stage from the cathode of the first video stage, and a positive-going composite signal is applied to the plate of the noise inverter from the plate of the first video stage.

When the inverter grid bias is adjusted correctly by the noise threshold control, the stage will be held at cutoff when the normal peak-to-peak composite signal is applied to the cathode, thus preventing the negative-going signal from reaching the plate. However, noise pulses greater in amplitude than the sync tips will drive the cathode sufficiently negative to cause conduction. Therefore, the noise pulses are amplified and

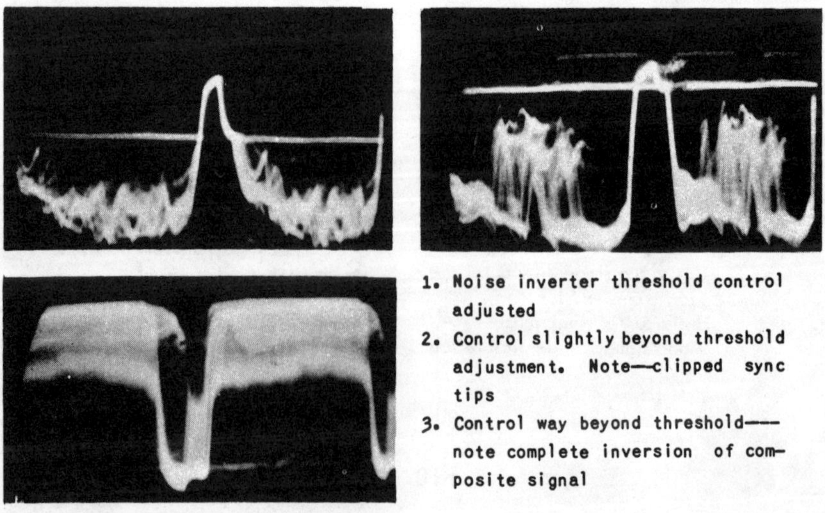

1. Noise inverter threshold control adjusted
2. Control slightly beyond threshold adjustment. Note—clipped sync tips
3. Control way beyond threshold—note complete inversion of composite signal

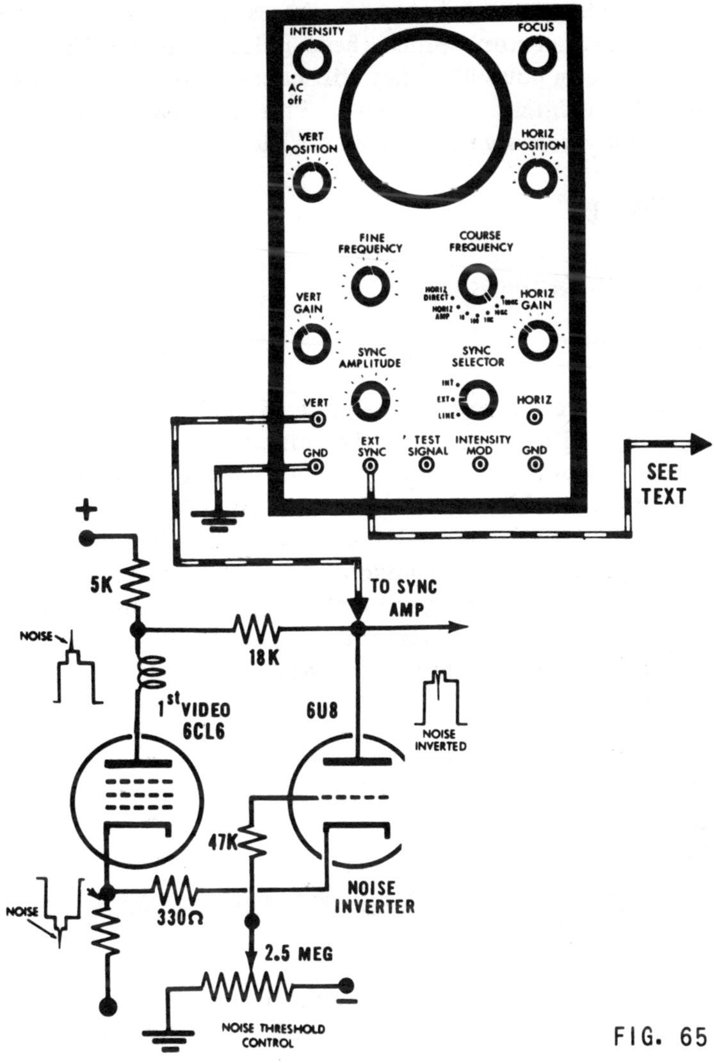

FIG. 65

appear at the plate of the inverter as strong negative pulses. The large negative plate pulses overcancel the positive noise pulses in the positive-going composite signal. In fact, the negative pulses are so strong that they actually invert the noise pulses. The positive-going composite signal with inverted noise pulses is then applied through a coupling capacitor to the grid of the first sync amplifier.

If the noise threshold control is not biasing the inverter sufficiently, the stage will have a passive function, allowing negative-going sync pulses from the cathode to appear at the plate of the inverter. Since these pulses are amplified and are negative in polarity, they will cancel, or in some cases invert, the original sync pulses. This will result in the loss of sync and excessive tearing of the picture.

Procedure

Equipment Required:

Oscilloscope

Connect an oscilloscope to the plate of the inverter stage and tune in a local channel. Turn the noise threshold control fully counterclockwise, then slowly rotate the control clockwise until it affects the positive-going sync tip. The oscillograms show normal and abnormal conditions. The pictures were taken while adjusting the noise control. In setting up the oscilloscope use external sync by clipping the external sync lead to the insulation of the horizontal yoke lead. Adjust the sweep for two horizontal sync pulses and the vertical gain for a 2-inch peak-to-peak pattern.

66

Checking Sweep and Marker Generators

It is advisable to make a periodic check of the sweep and marker generator with a known standard. A test setup using a 4.5-MHz IF transformer is shown in Fig. 66.

Procedure

Equipment Required:

Wideband oscilloscope
Demodulator probe (high-impedance)
Sweep and marker generator
4.5-MHz IF transformer

Connect the equipment as shown in Fig. 66. The shielded cable from the generator should be terminated by an appropriate resistance. Switch on the test equipment, allow a warmup period, and adjust the sweep and marker frequencies to 4.5 MHz (sweep width 2 MHz with 4.5 MHz center). The marker pip should be riding on the peak of the response curve, providing the IF transformer has been pre-set at this frequency by a crystal-controlled marker (see oscillogram).

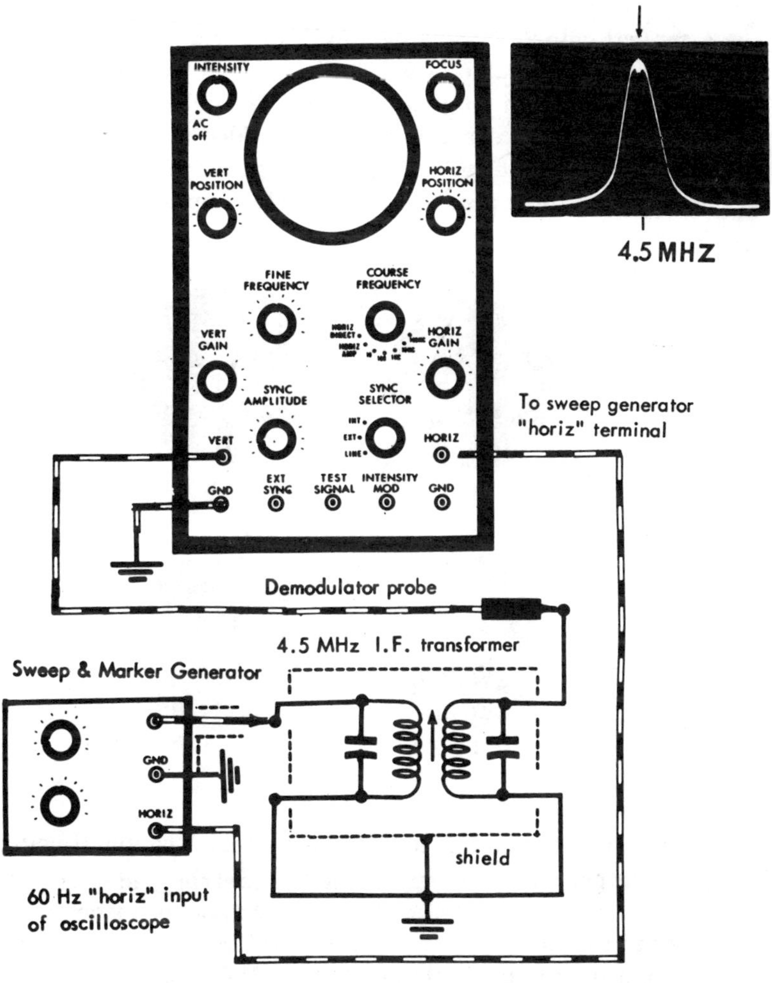

FIG. 66

67

Blanking Amplifier Waveforms

A typical color TV blanking amplifier circuit is shown in Fig. 67. A flyback pulse is fed to the input at position 1 and appears inverted in the plate circuit. The pulse at position 2 is applied to the screen grid of the last video stage and drives it into cutoff, thus eliminating burst signal interference. In the absence of this pulse the burst signal appears as a vertical

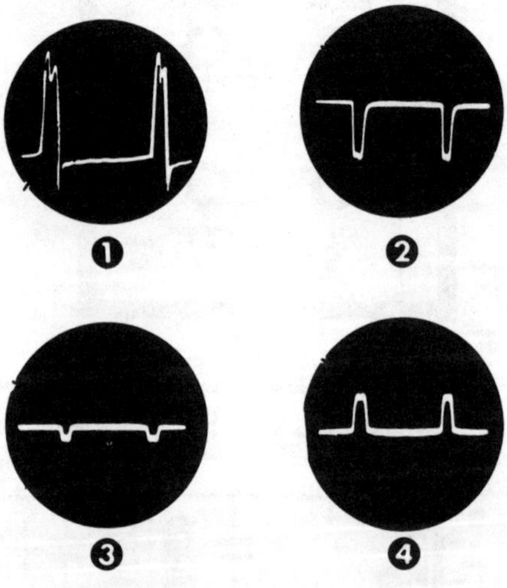

color stripe on the left side of the raster. The small negative pulse at position 3 is applied to the control grid of the killer stage and a positive pulse from the cathode (position 4) is applied to the color control. (See oscillograms corresponding to position numbers above.) Any slight malfunction of the blanking amplifier would effect both color and black-and-white reception.

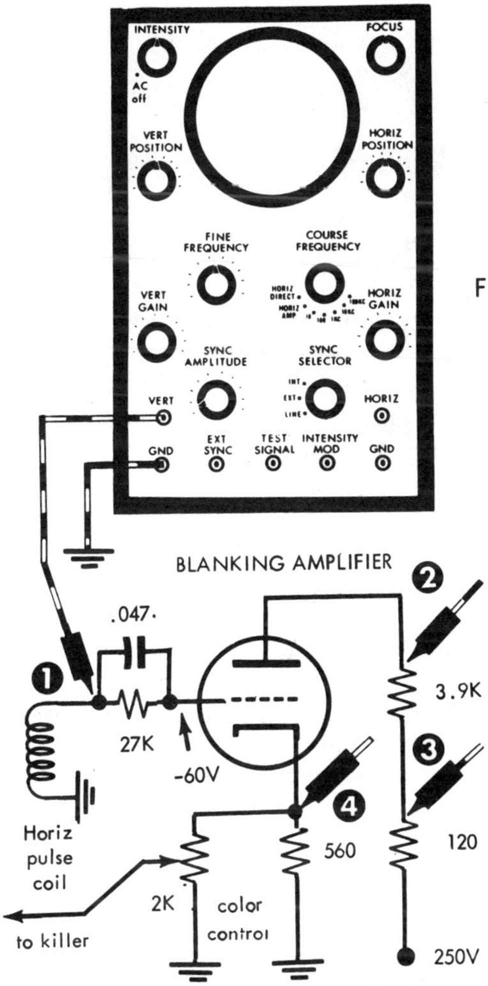

FIG. 67

Procedure

Equipment Required:

Wideband oscilloscope

Each pulse should be present at the points specified above and must measure up to the manufacturer's specifications. Also, the pulse distribution points should be checked; in this case, the screen grid of the last video, the killer control grid, and the color control.

68

Vertical Blanking Pulse

A strong vertical retrace blanking pulse supplied by the vertical deflection amplifier is applied to the three cathodes of the tri-color picture tube and timed to blank out the vertical retrace. Although the vertical deflection system is operating satisfactorily, the coupling circuit to the picture tube should be checked for a defective coupling capacitor when retrace lines appear. (See Fig. 68.)

Procedure

Equipment Required:

Wideband oscilloscope, low-capacity probe

Check for the presence of the blanking pulse at the point of origin in the vertical deflection amplifier. The pulse should also appear as a strong positive pulse at each CRT cathode. As shown in the oscillograms, the pulse appears to peak well above the horizontal sync tip level. In color TV the blanking pulses have to cut off three cathode-ray beams, hence its rugged appearance.

❶ ❷

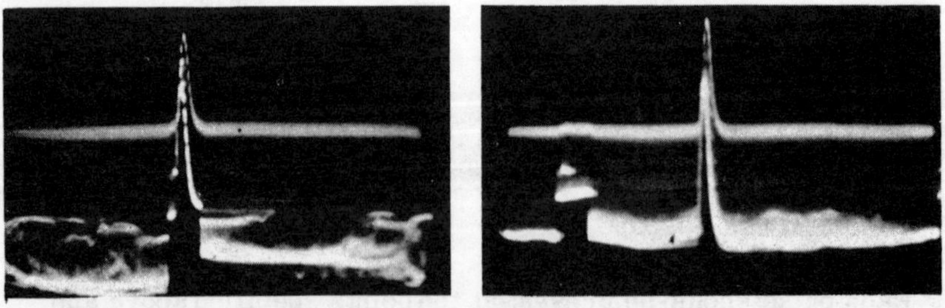

1. Sync and blanking pulses synchronized
2. Blanking out of sync. Notice that the vertical sync pulse is far below the horizontal sync pulse line, indicating poor low-frequency response. This is possible when using a low-capacity test probe.

FIG. 68

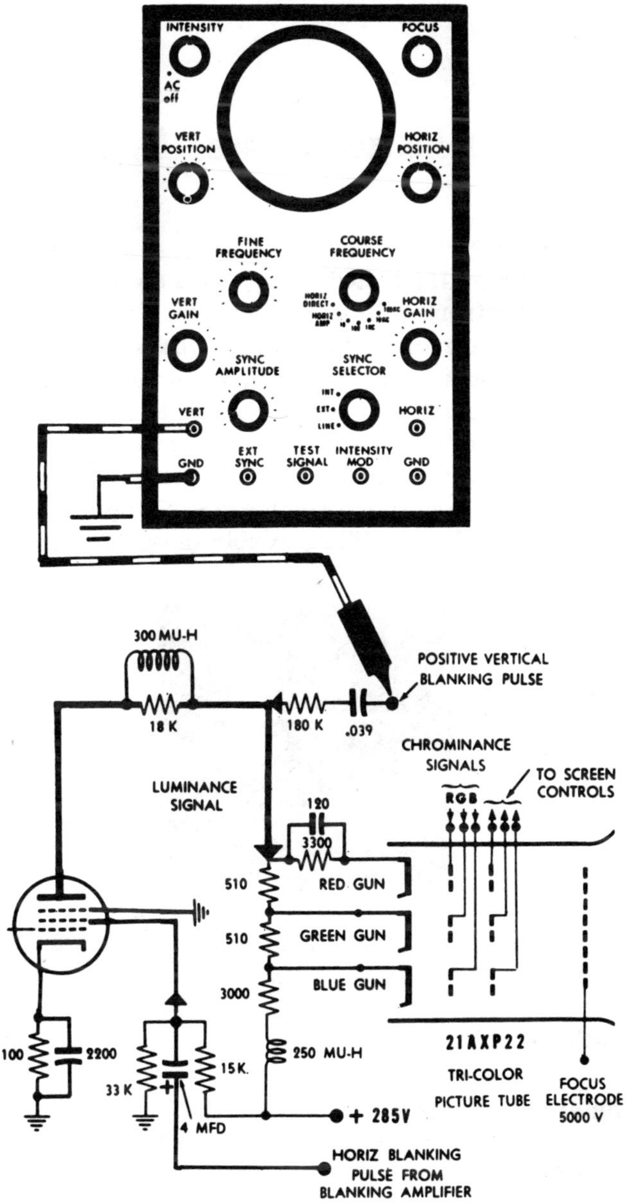
INTENSITY
FOCUS
AC
off
VERT
POSITION
HORIZ
POSITION
FINE
FREQUENCY
COURSE
FREQUENCY
VERT
GAIN
HORIZ
GAIN
SYNC
AMPLITUDE
SYNC
SELECTOR
VERT
HORIZ
GND
EXT
SYNC
TEST
SIGNAL
INTENSITY
MOD
GND
300 MU-H
POSITIVE VERTICAL
BLANKING PULSE
18 K
180 K
.039
CHROMINANCE
SIGNALS
TO SCREEN
CONTROLS
LUMINANCE
SIGNAL
120
RGB
3300
510
RED GUN
510
GREEN GUN
3000
BLUE GUN
100
2200
33 K
15 K.
250 MU-H
21AXP22
TRI-COLOR
PICTURE TUBE
FOCUS
ELECTRODE
5000 V
4 MFD
+ 285V
HORIZ BLANKING
PULSE FROM
BLANKING AMPLIFIER

69

Checking Horizontal Phase Detector

The phase detector illustrated in Fig. 64 is an important juncture in a TV horizontal deflection system. Its purpose is to compare the timing of an integrated flyback pulse with the horizontal sync pulse. For example, when the two pulses are in step, the phase detector output voltage is zero. However, should any phase difference occur between the received sync pulses and the horizontal oscillator frequency, the phase detector transmits a correction voltage to speed up or slow down the horizontal oscillator. When trouble in the horizontal hold

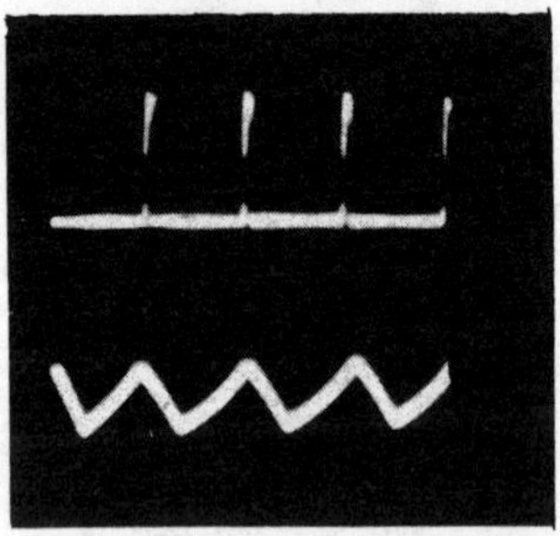

is experienced, a check for the presence of the two pulses is in order.

Procedure

Equipment Required:

Oscilloscope

Set up equipment as illustrated in Fig. 69. Adjust oscilloscope sweep rate to 10 kHz, and the fine frequency control to display two pulses. If both pulses are present, check their amplitudes with the manufacturer's specifications. See the oscillograms for normal waveforms.

FIG. 69

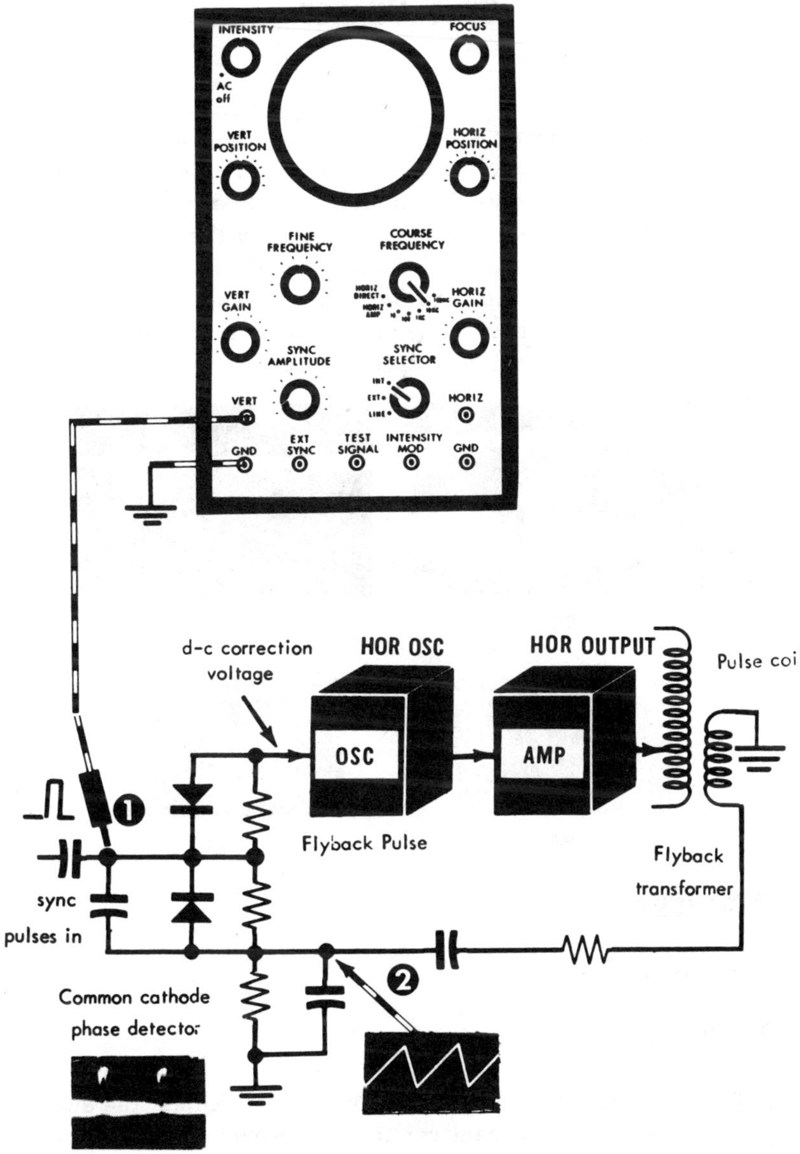

70

Sync Clipper

A typical sync clipper stage is shown in Fig. 70. A sample of the composite video signal is taken from the video stage and applied to the control grid of the clipper stage. In the process the sync pulses are separated from the video portion of the composite signal, and the separated sync pulses, both vertical and horizontal, are fed to their respective circuits for further processing. These circuits should be checked when horizontal and vertical sync troubles occur.

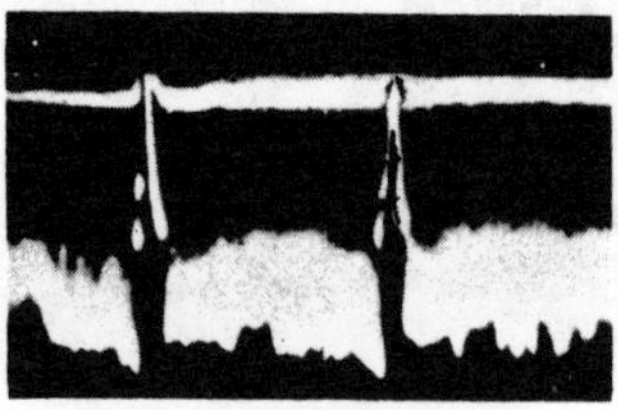

❶

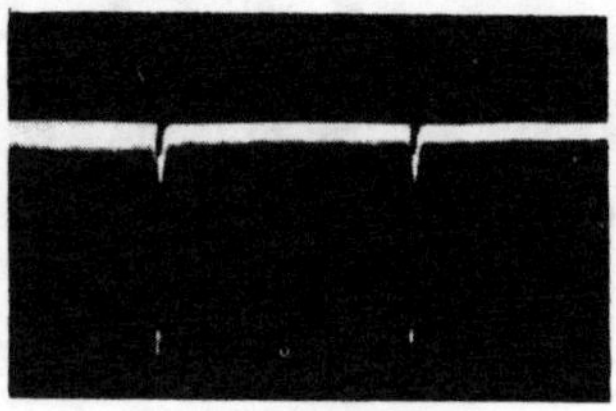

❷

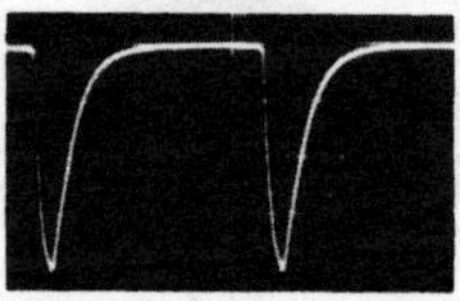

❸

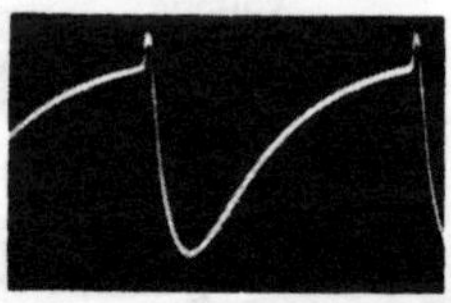

❹

Procedure

Equipment Required:

Oscilloscope

Set up the oscilloscope as shown in Fig. 70. Adjust the sweep rate to 30 Hz and connect vertical input probe to the clipper control grid. Tune in a local channel and observe the composite signal (see Oscillogram 1). Connect the probe to the plate of the clipper and measure the separated vertical sync pulses (Oscillogram 2). When probing the plate of the clipper, switch the sweep rate to 7875 Hz to assure the presence of horizontal sync pulses (Oscillogram 3). Return the sweep rate to 30 Hz and connect the probe to the output of the vertical integrator; integrated vertical sync pulses should appear here (Oscillogram 4). Refer to the manufacturer's specifications for peak-to-peak values.

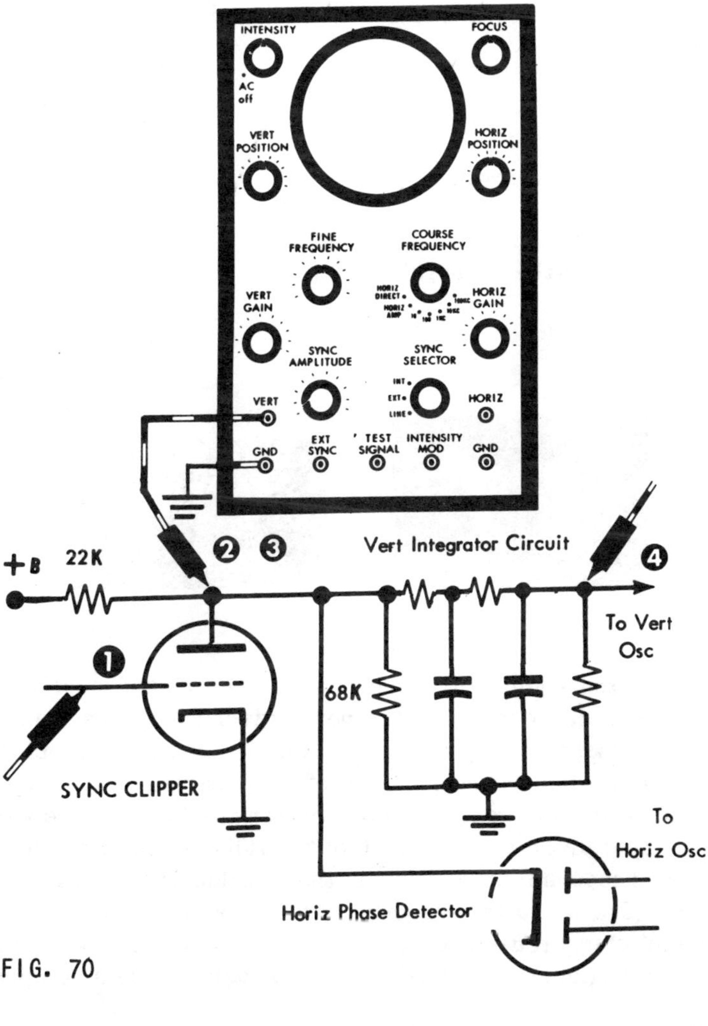

FIG. 70

71

Last IF Amplifier Response Curve (Solid-State)

The procedure for checking the response curve of a transistor video IF amplifier is similar to that used for a vacuum tube amplifier.

Procedure

Equipment Required:

Wideband oscilloscope
Sweep and marker generator

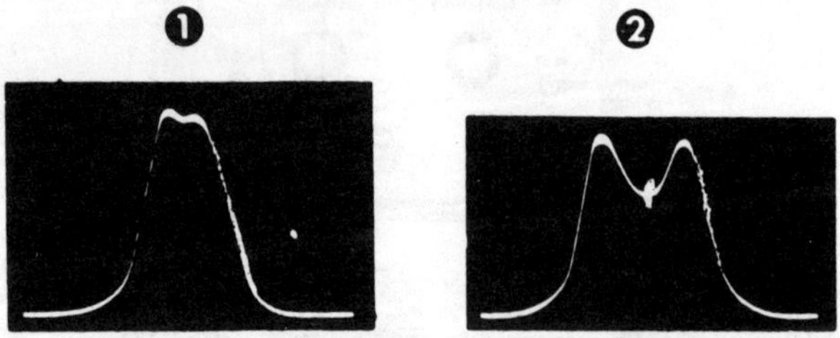

Connect the equipment as shown in Fig. 71 and adjust the sweep generator to the IF frequency given in manufacturer's specifications, and at a sweep width of 5 MHz. Connect the horizontal phased sweep output of the generator to the horizontal input of the oscilloscope. Attach the oscilloscope vertical input probe to the output of the video detector and align as per response curve shown in the manufacturer's alignment data. (See a typical response curve in oscillogram.) In the case of staggered-tuned circuits, each stage must be peaked as per alignment instructions. Oscillogram 1 is a compromise with that of Oscillogram 2. The center frequency dip should not exceed more than 30% of the total amplitude, despite its broader response.

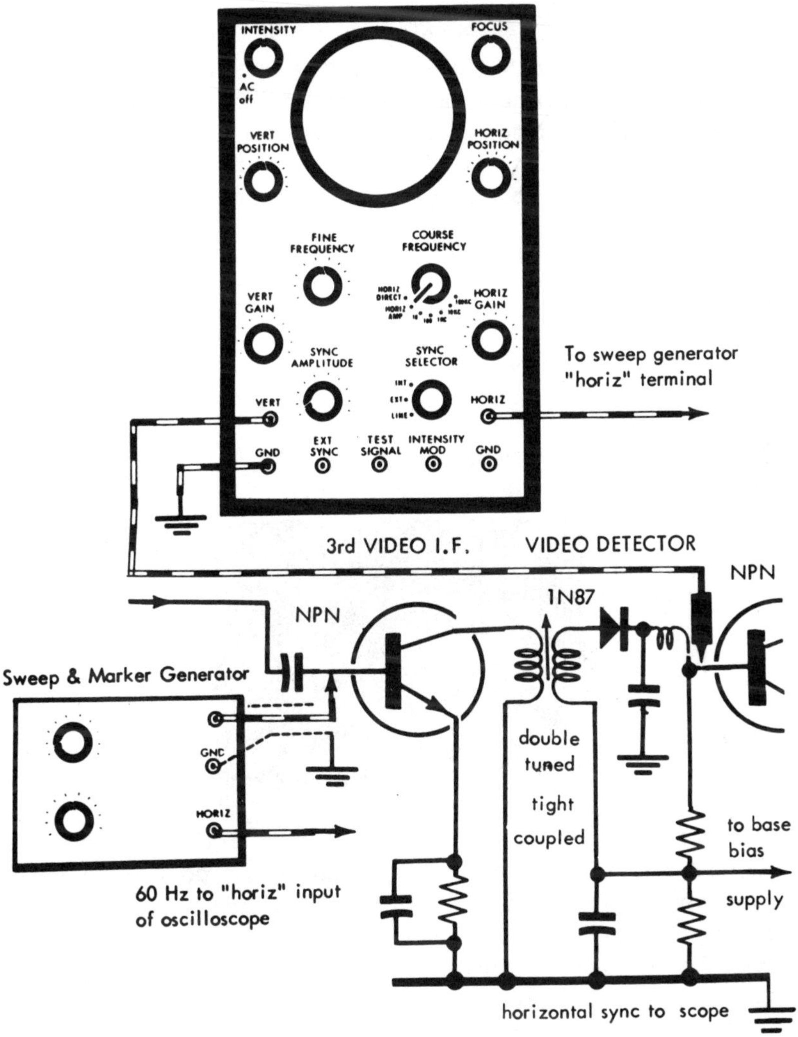

FIG. 71

72

Last IF Amplifier Response Curve (Vacuum Tube)

When the IF section needs alignment it is a good practice to start the procedure at the last IF stage and proceed toward the mixer. It is common practice not to disturb the constants of a circuit by connecting test equipment. If a demodulator probe is used it must be a high-impedance type; therefore, it is advisable to connect the vertical input to the second detector output. See Fig. 72.

Procedure

Equipment Required:

Oscilloscope
Sweep and marker generator

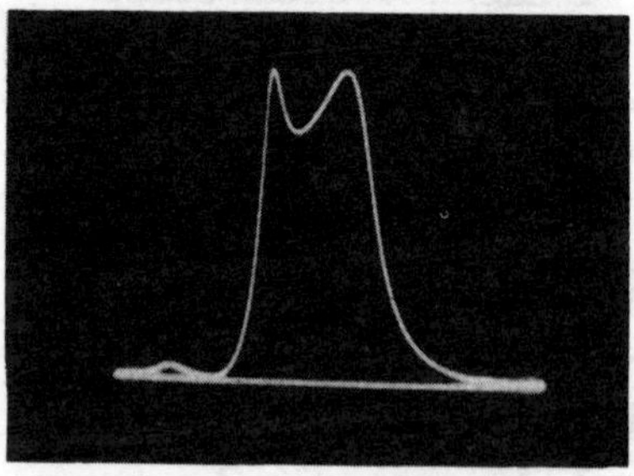

Connect the sweep generator to any point in the IF section, using a small capacitor to reduce loading. A point to be remembered: When a tuned IF circuit precedes the sweep signal input point, a dip in the response curve may appear at this frequency, due to absorption. Fortunately, this can happen only at the fundamental frequency—harmonics are ruled out. In the oscillogram, notice the absorption dip caused by the tuned circuit preceding the stage being aligned. When this circuit was detuned the dip disappeared.

Connect the generator output lead to the control grid and

align the plate coil; do not attempt to align the grid coil to which you attach the output lead. A small capacitor should be in series to block the DC operating voltage associated with the amplifier, especially when connected to the plate; in this case the terminating resistor could burn out since one end of the output cable is grounded. Do not use a low-capacity probe because the response curve being observed is 60 Hz. While aligning stages that are AGC controlled, remove the AGC lead and replace with a small battery voltage, controlled by a potentiometer.

FIG. 72

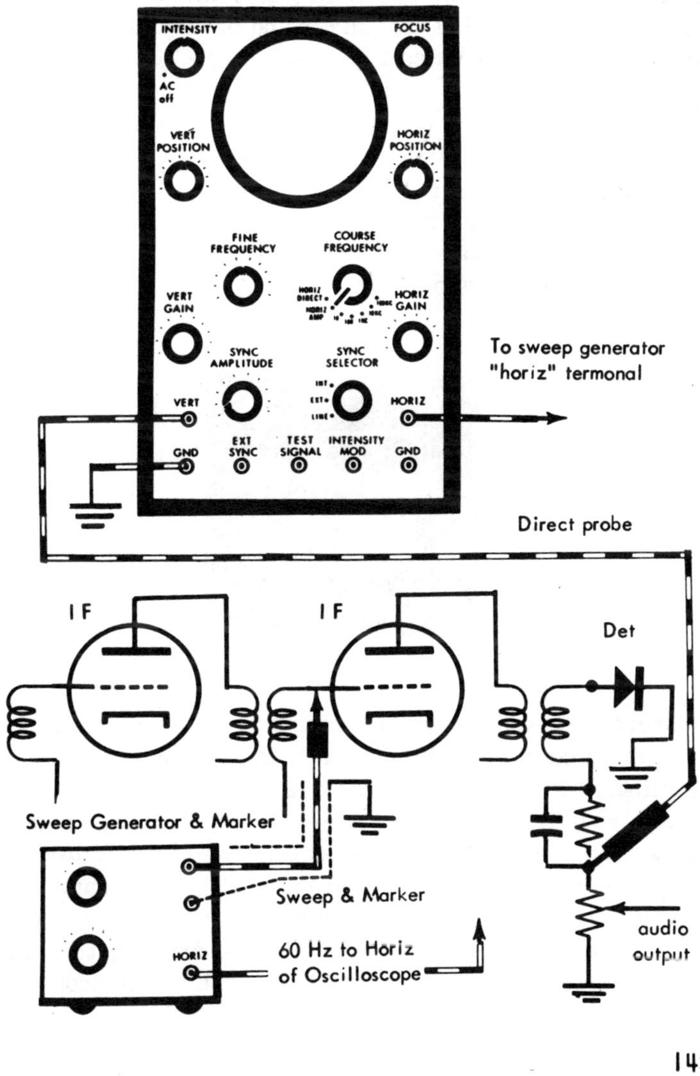

73 Overall IF Amplifier Sweep Alignment

An overall IF response curve may be obtained by introducing the alignment signal at the mixer and observing the output at the second detector. See Fig. 73.

Procedure

Equipment Required:

Wideband oscilloscope
Sweep and marker generator

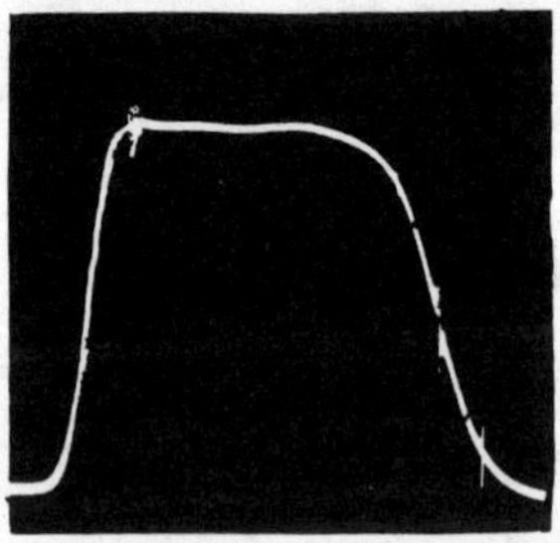

Connect the sweep generator output to an ungrounded shield over the mixer tube. Remember, the output cable must be terminated with an appropriate resistance, per the manufacturer's specifications; also, the cable shield must be grounded to the receiver chassis. Adjust the sweep generator to the frequency range conforming with the alignment data. If the marker is from a separate generator, connect its output to some convenient point in the IF channel prior to the second detector. Connect the vertical input of the oscilloscope to the output of the second detector and switch the station selector to an unused channel. A crystal-controlled marker generator is advisable, and is a must in color TV alignment.

FIG. 73

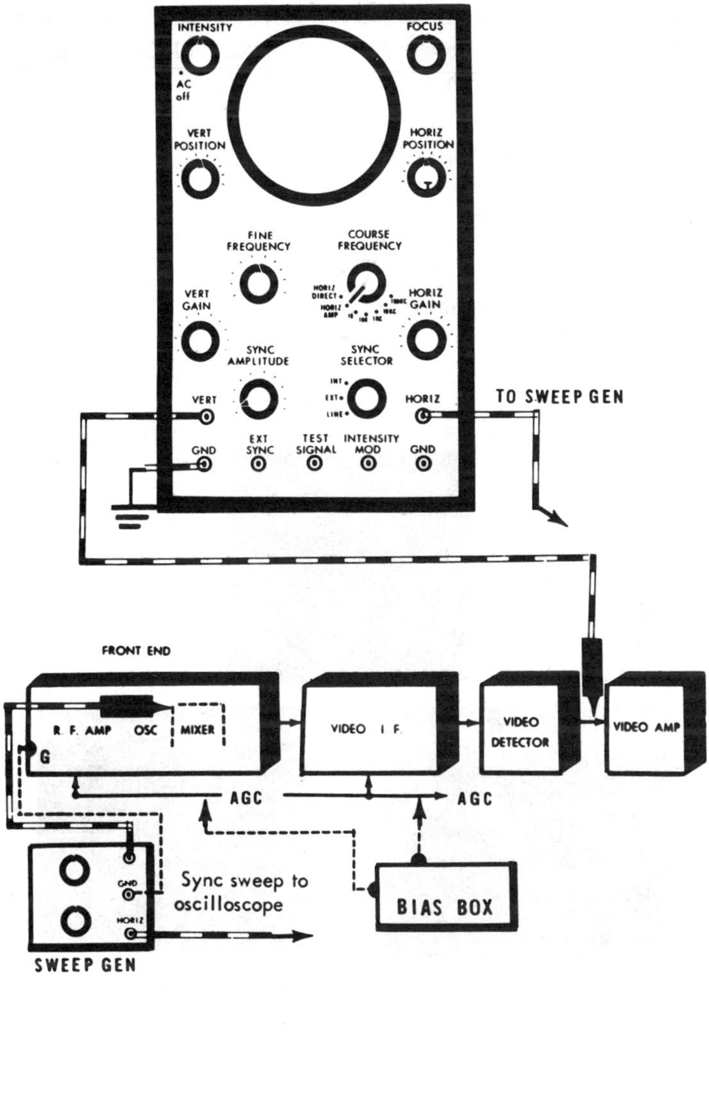

74

Sound Detector Alignment

To align the sound detector in a receiver using intercarrier sound, the sweep dial is set at 4.5 MHz. For accuracy a 4.5-MHz crystal should be used. A crystal jack is usually provided by the manufacturer.

Procedure

Equipment Required:

Oscilloscope
Sweep and marker generator
4.5-MHz crystal
Direct probe

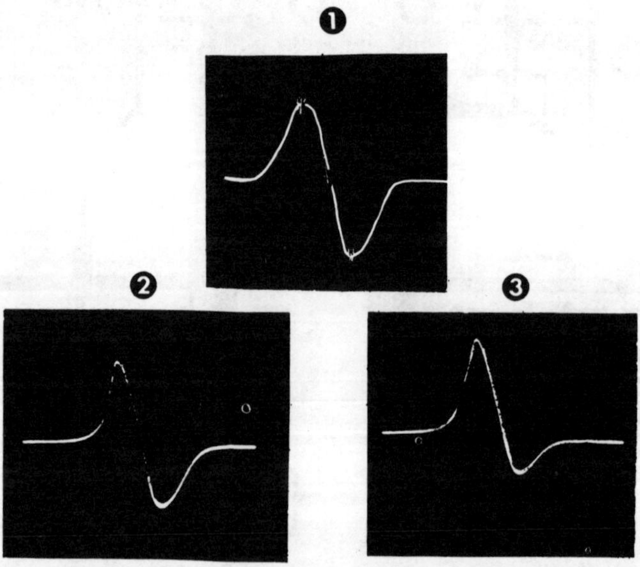

1. S curve in ideal symmetry with the markers
2. Incorrect S curve; left peak suppressed
3. Incorrect S curve; both peaks distorted

Set up equipment as shown in Fig. 74 and adjust the phasing control for a single trace. Set the sweep frequency at 4.5 MHz and adjust the sweep width to accommodate the trace. The two peaks of the S curve should be about 250 kHz apart. Align transformer T3 to provide a symmetrical S curve as illustrated.

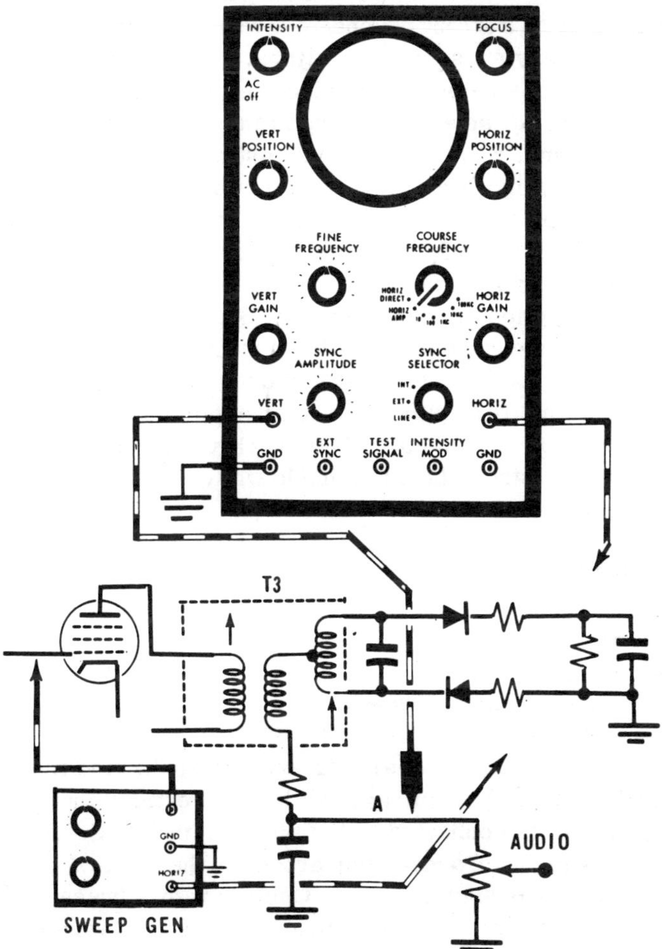

FIG. 74

75 Keyed AGC Waveforms

AGC provides a constant RF and IF gain within its range when switching from a strong channel to a weaker channel and under fading or varying signal conditions. To accomplish this a special stage is designed to select and sample the horizontal sync pulses. A keyed AGC stage is shown in Fig. 75. This stage is normally cutoff by AGC adjustment R2, a preset control that may need occasional readjustment.

A positive pulse initiated in the flyback transformer is applied to the plate of the AGC tube and is timed to gate the tube into conduction during horizontal retrace. A sample of the video signal is applied to the control grid and the sequence of the keying pulses causes the tube to conduct during retrace time and amplify the horizontal sync pulses applied to the control grid. The amplified pulses are then applied to the AGC filter capacitors where a DC control voltage is produced. A periodic check of this stage should be made while the chassis is being serviced.

Procedure

Equipment Required:

Oscilloscope, direct probe

Set the sweep rate to 7875 Hz and adjust the fine frequency control to display 2 pulses, while the vertical input of the oscilloscope is connected to the plate of the AGC tube. See Oscillogram 2. Connect the probe to the control grid and check the video signal against Oscillogram 1 (sweep rate 30 cycles). Finally, connect the probe to the cathode and observe the pulses present here, an indication the stage is conducting. Do not disturb the AGC adjustment, unless the waveforms in the set do not match those shown in the oscillograms.

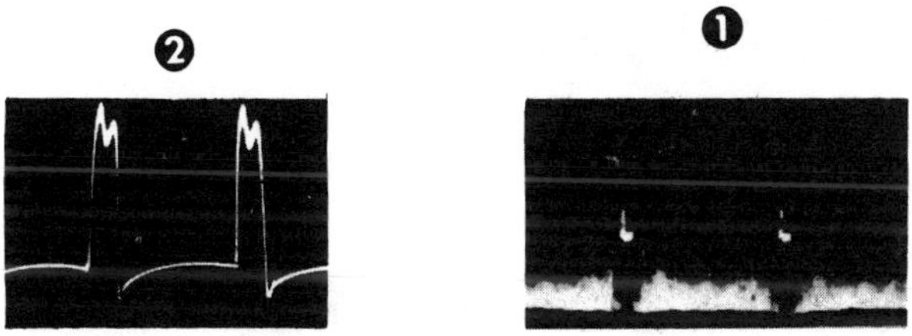
2
1

FIG. 75

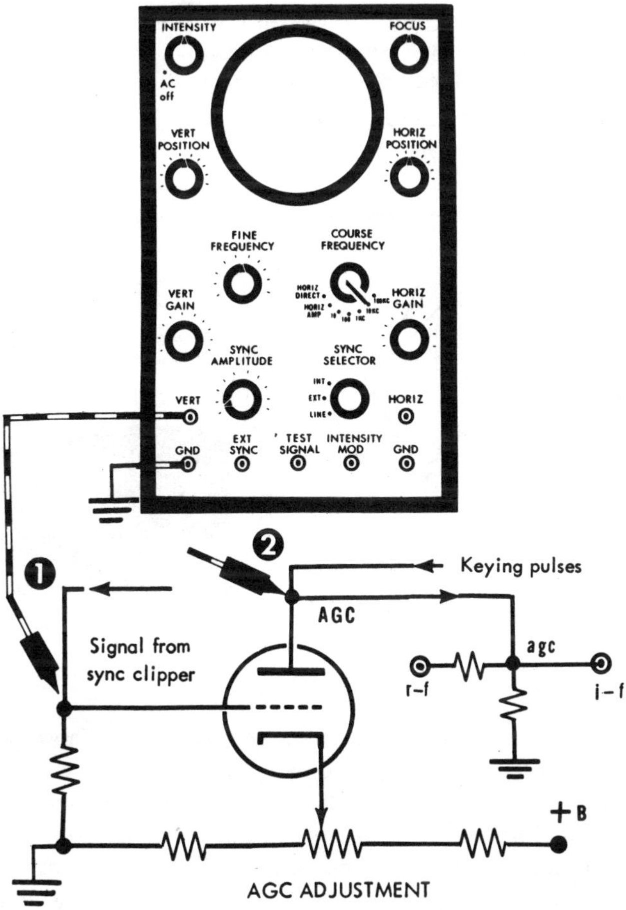
INTENSITY
FOCUS
AC
off
VERT POSITION
HORIZ POSITION
FINE FREQUENCY
COURSE FREQUENCY
VERT GAIN
HORIZ GAIN
SYNC AMPLITUDE
SYNC SELECTOR
VERT
HORIZ
GND
EXT SYNC
TEST SIGNAL
INTENSITY MOD
GND
1
2
Keying pulses
AGC
Signal from sync clipper
agc
r-f
i-f
+B
AGC ADJUSTMENT

76
77
78

Horizontal Sync Systems

Horizontal sync systems vary with the type and manufacture of the receiver. Here are the three most often encountered. Fig. 76 illustrates the "synchrolock" circuit. The received sync pulse and a sample of the flyback pulse are applied to this circuit. The timing of both pulses is shown in the oscillogram for normal synchrolock action and any shift of the flyback pulse will develop a correction voltage to speed up or slow down the horizontal oscillator. Another horizontal sync

FIG. 76

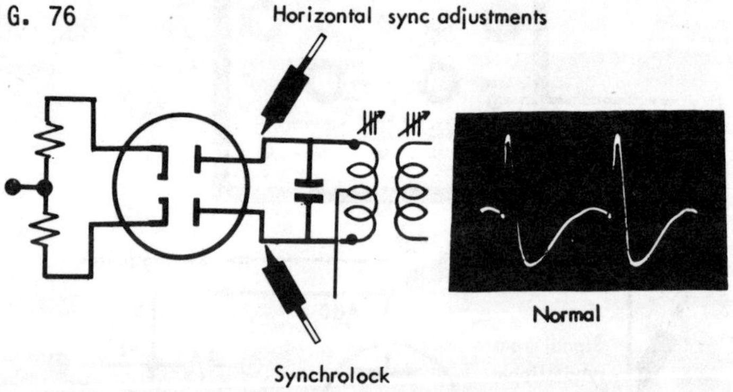

control system, called the "synchroguide," is shown in Fig. 77. The coil is adjusted (tuned) until the sine wave peaks have the same amplitude. The circuit in Fig. 78 also employs the flywheel action of a tank circuit in the horizontal multivibrator plate circuit, appropriately called a "stabilizing coil." Notice the position of the injected flyback pulse on the sine wave. (All the above oscillograms were taken with a sweep rate of 7875 Hz.)

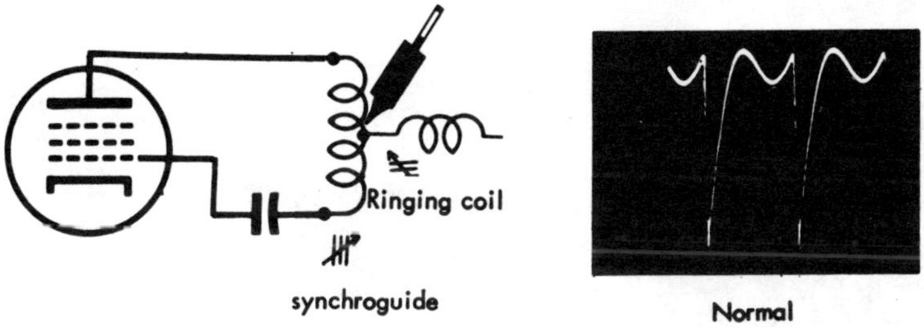

FIG. 77

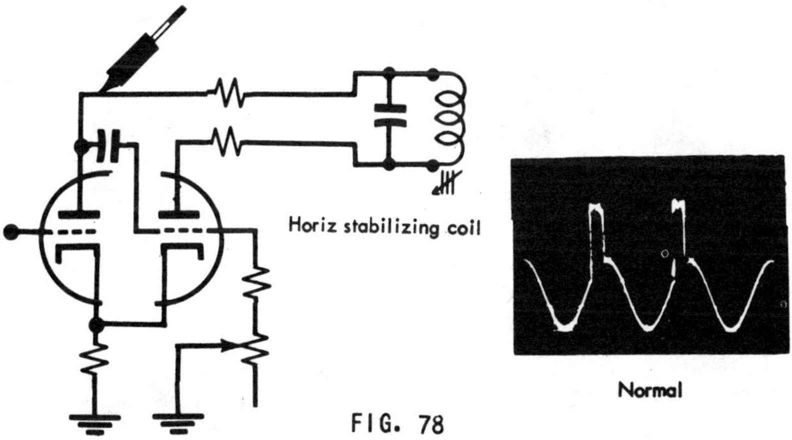

FIG. 78

79

Distribution of Burst and Chrominance Signals

The distribution of the luminance, chrominance, and burst signals is shown graphically in Fig. 79. The luminance signal is applied to the second video stage and supplies the brightness information to the three picture tube cathodes. When the vertical input of the oscilloscope is connected to the input and output of the second video stage, video signal amplification may be evaluated, using a standard color bar generator and an oscilloscope sweep rate of 7875 Hz.

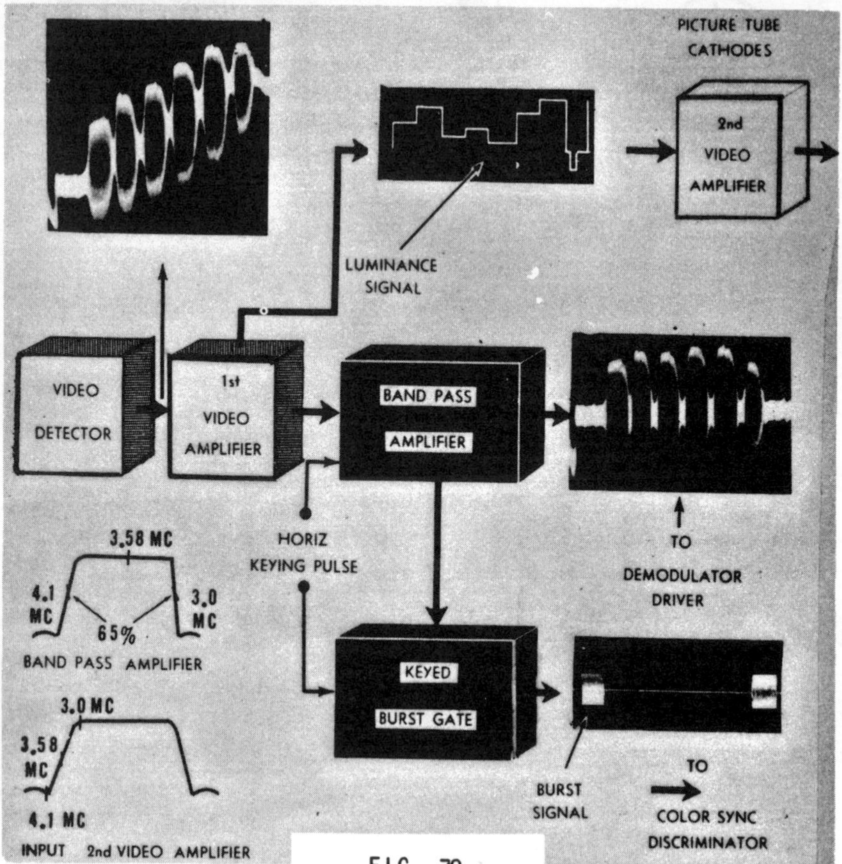

FIG. 79

The burst signal is applied to the color sync discriminator where it is compared in phase to the color subcarrier oscillator. The output of the discriminator then controls the reactance tube which represents a capacitive reactance across the tank circuit of the oscillator.

The chrominance signal is applied to a demodulator driver which in turn drives a pair of demodulators for the recovery of the color-difference signals. These signals are then applied to their respective amplifiers and finally to the control grids of the picture tube.

80

Color Gating Pulse

In a color receiver a flyback pulse is applied to the blanking amplifier (see Fig. 80). The blanker output feeds a small negative pulse to the color "killer" control grid, resulting in a positive pulse at the "killer" plate. The positive pulse is applied to the control grid of the bandpass amplifier which is gated into conduction during burst signal time. If a burst signal is present it functions as a pilot signal and passes through the bandpass amplifier where it is amplified and applied to a discriminator, rectified and filtered, resulting in a negative voltage at the killer control grid. The negative voltage causes this stage to cut off and remain in this condition during a color transmission.

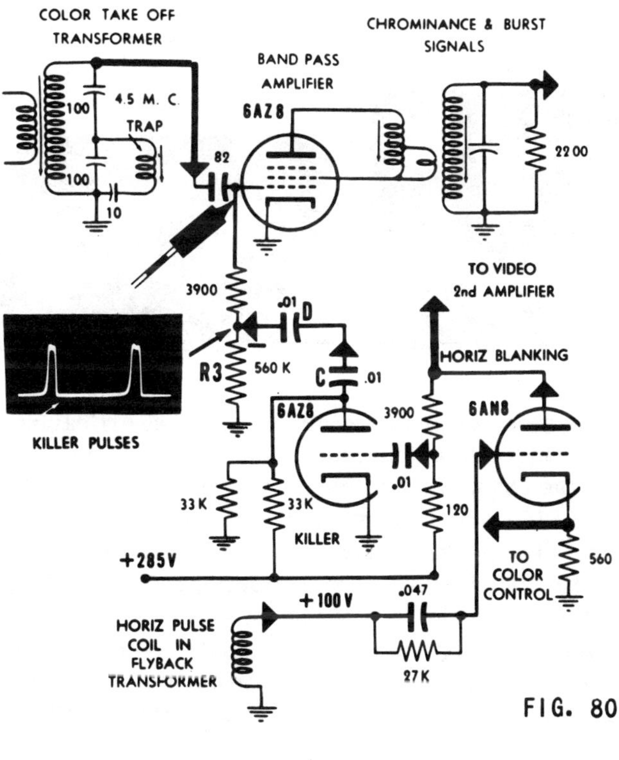

FIG. 80

81

Flyback Transformer Pulse Coil Waveforms

A horizontal flyback transformer can be checked by viewing the pulse delivered by the coil, and the high voltage by a proximity test. (This test is for black-and-white receivers only.)

Procedure

Equipment Required:

Oscilloscope

❶

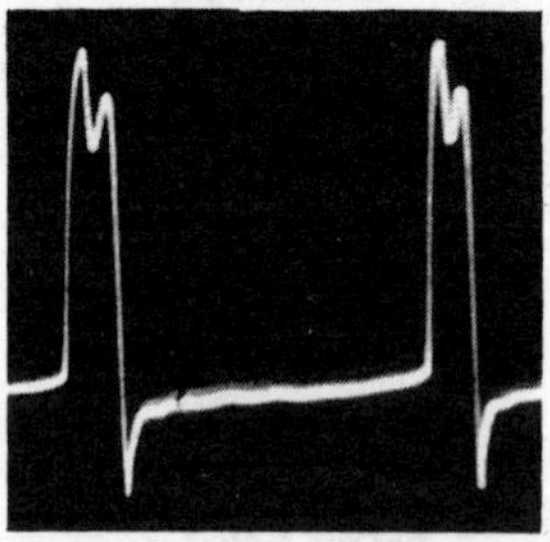

❸

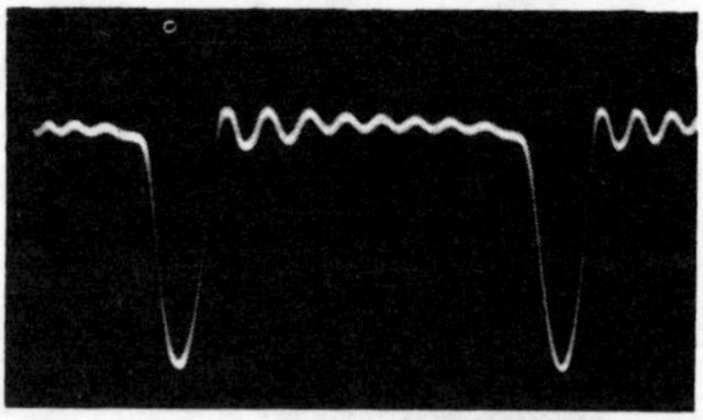

With the receiver switched off connect the vertical input of the oscilloscope to the output lead of the pulse coil and the ground lead to the chassis. Switch on the receiver. Set the sweep rate to 7875 Hz and adjust the fine frequency until two pulses appear. The peak-to-peak value of these pulses should conform with manufacturer's specifications. While making this test, checks should be made at the pulse distribution points.

For the proximity test, switch off the receiver and clip a plastic clothespin to the insulation on the horizontal output tube plate lead and clip the oscilloscope vertical input to the end

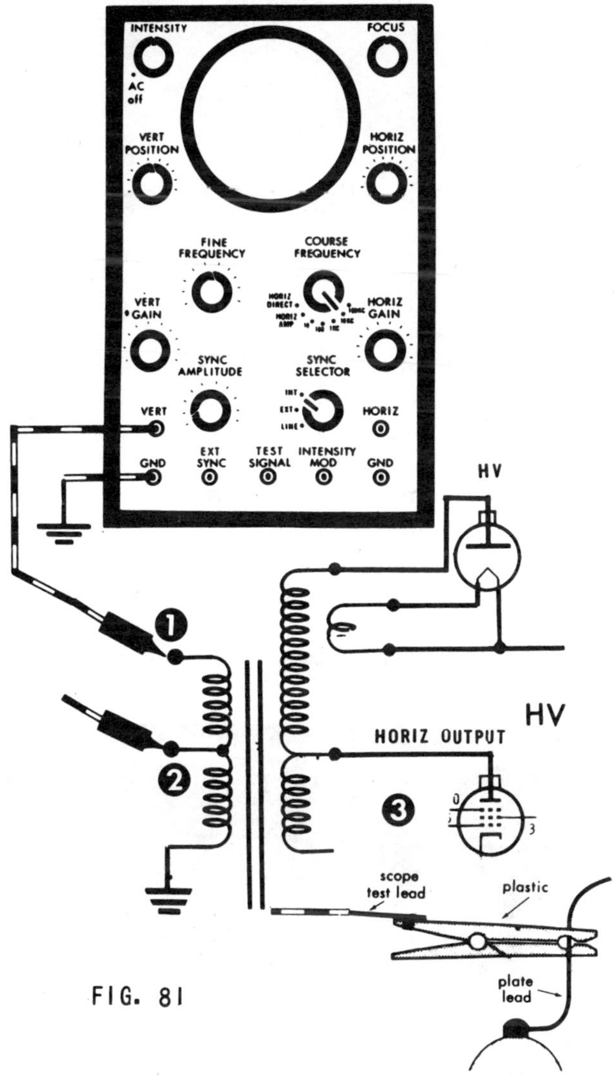

FIG. 81

Illustrating the proximity test.

of the pin as illustrated in Fig. 81. The lead will be about two inches removed from the high-voltage plate lead but close enough to pick up the flyback oscillations. Do not switch on receiver until the test set up is complete, and switch off receiver after making the observation. All these precautions are given to avoid the discomfort of receiving an electrical shock. Advise all personnel accordingly.

82 Color TV Alignment Notes

Broadband RF and IF amplifier response in color receivers is of utmost importance. Although a 4-MHz video bandwidth is the accepted standard for black-and-white receivers it is seldom used. In fact, a recent check of black-and-white receivers showed that the average bandwidth of ten different makes was approximately 3.4 MHz. The IF bandwidth of a color receiver must extend to a 4.2 MHz, if proper amplification of the upper sidebands of the color subcarrier is to be achieved. It will be noted that the flat portion of the response curve extends almost to the sound carrier location.

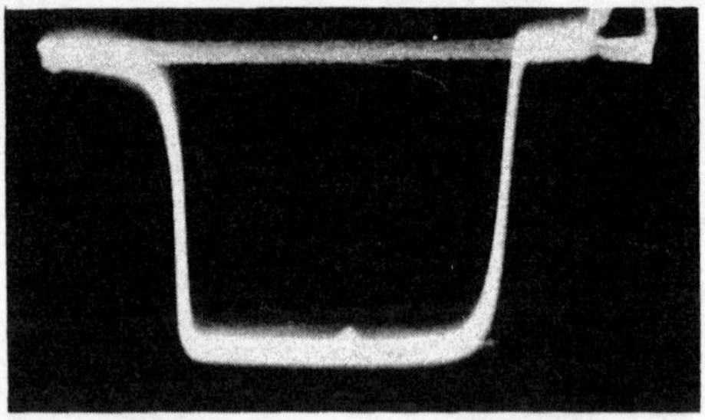

Actually, the color subcarrier is only 920 kHz removed from the sound carrier. Also notice that the sound carrier is sharply attenuated; therefore, alignment in this region is critical. If the flat portion of the curve does not extend out far enough to include the upper subcarrier sideband, poor color reproduction will result, and in severe cases the color information will be lost entirely.

With this close bandwidth tolerance, the technician is confronted with more critical alignment problems than he experienced with black and white. For this reason the signal generator must have precise marker indications and be capable

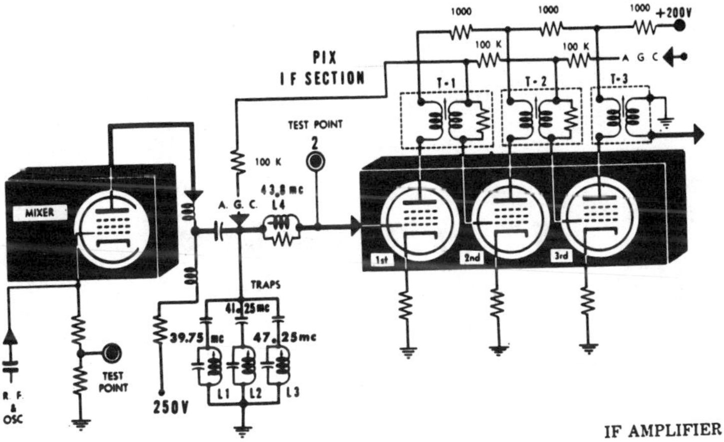

FIG. 82

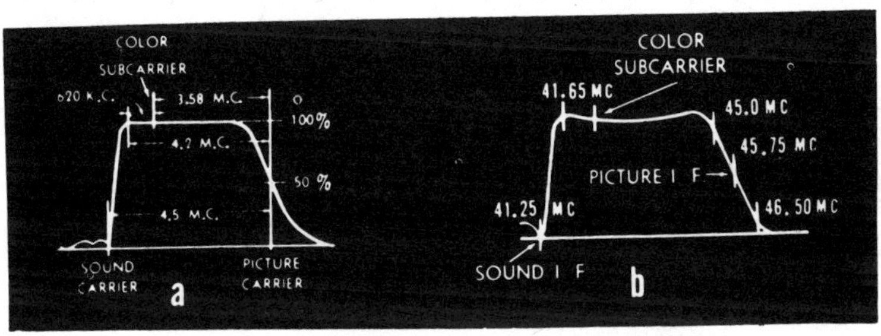

of providing two frequencies that are only a fraction of a MHz apart, one peaking at 41.65 MHz and the other with maximum attenuation at 41.25 MHz. These frequencies must be exact and not approximate, and the sweep generator must be linear over its entire range. Should there be any doubt as to the accuracy of test equipment, color receiver alignment should not be attempted. Crystal-controlled frequencies should be used so as to keep the alignment procedure organized. The response curve in the oscillogram is inverted on the graticule; some oscilloscopes have a reversing switch, but the position of the response curve is solely the preference of the operator. (The noise in the oscillogram case was due to local conditions in the laboratory and not in the receiver.)

83

TV Tuner Response Curves

The bandwidth of a color receiver tuner is much wider and more critical than that required in a black and white receiver. The correct tuner response curve is shown in the oscillograms in Fig. 83. A casual touch up in tuner alignment is not advisable unless the technician can observe the curve while adjusting. Also, it must be remembered that the local oscillator or mixer stage is not involved in tuner alignment procedure. However, during an overall alignment the tuner stages should be included. Be sure to disable the automatic fine tuning circuit, if used, and do not disturb tuner shielding except where absolutely required to permit access to adjustments.

Procedure

Equipment Required:

Wideband oscilloscope
Sweep and marker generator
VTVM
Bias supply (2v)

Do not attempt tuner alignment unless crystal-controlled markers are available, and then follow precisely the manufacturer's instructions. Connect the oscilloscope through an appropriate matching network to the tuner output and the sweep generator to the antenna terminals. Unless connected internally to the sweep, loosely couple the marker generator to the sweep cable. (The VTVM is used to measure the oscillator injection voltage.)

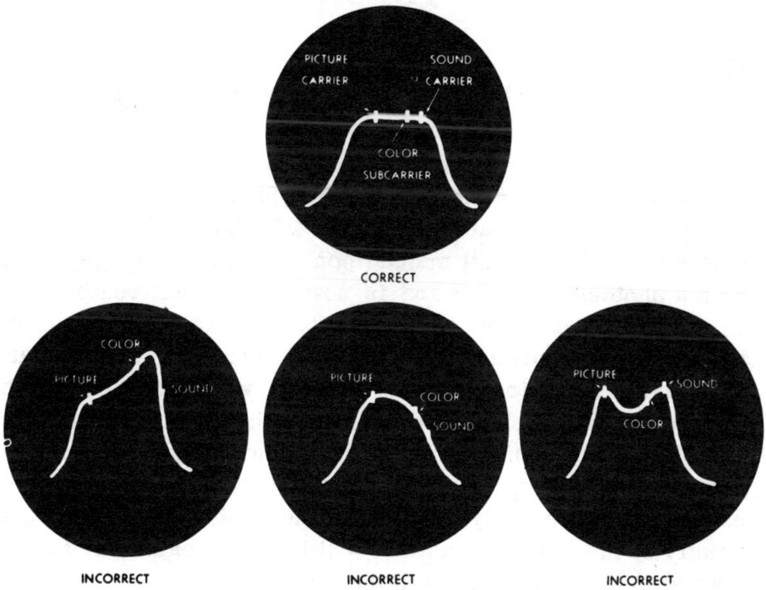
PICTURE CARRIER
SOUND CARRIER
COLOR SUBCARRIER
CORRECT
COLOR
PICTURE
SOUND
INCORRECT
PICTURE
COLOR
SOUND
INCORRECT
PICTURE
SOUND
COLOR
INCORRECT

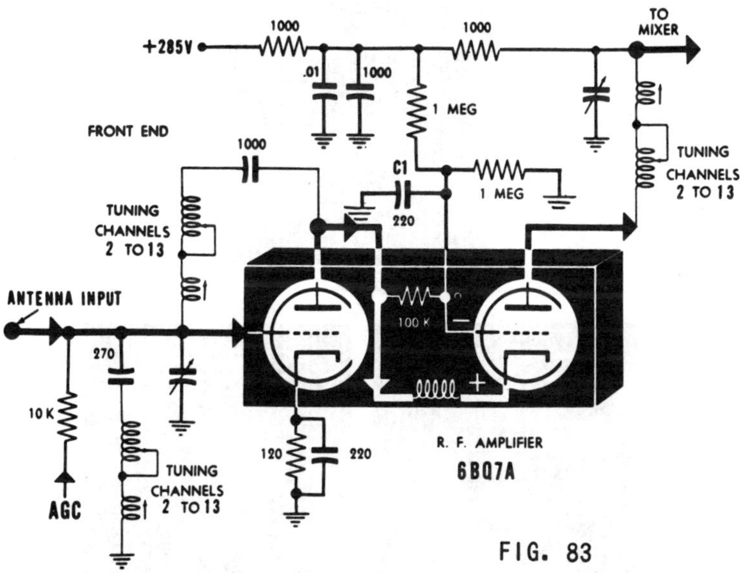
+285V
1000
1000
TO MIXER
.01
1000
1 MEG
FRONT END
1000
C1
1 MEG
220
TUNING CHANNELS 2 TO 13
TUNING CHANNELS 2 TO 13
ANTENNA INPUT
100 K
270
10 K
120
220
R. F. AMPLIFIER
6BQ7A
TUNING CHANNELS 2 TO 13
AGC

FIG. 83

84 Photoelectric Phase Control Curves

An SCR operating in conjunction with a photocell is illustrated in Fig. 84. The circuit is similar to a halfwave phase control; however, in this circuit manual control R2 has been replaced with a photoelectric cell and is now light sensitive, and the time constant of the gating circuit is varied by the amount of light striking the photocell. Shunting the photocell with a 2-watt 10K resistor will increase the sensitivity of the circuit. With this parallel resistance the discharge path will not exceed 10K; therefore, a slight change in resistance (dim light) of the photocell will reduce the time constant of the discharge path of C1 sufficiently to trigger the SCR. Then as the intensity of the light increases it will be accompanied by an increase in conduction through the load.

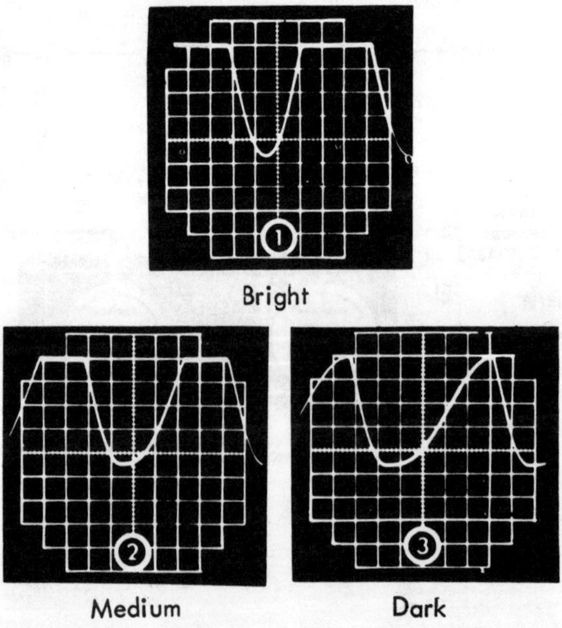

Bright

Medium Dark

Capacitor charge and discharge waveforms

A 100-volt 10-watt lamp connected across the load and located near the photocell will further increase the sensitivity of the circuit. For example, the flame from a match held three feet from the photocell will cause the 10-watt lamp to glow sufficiently to activate the photocell still further. This, in turn, increases the brightness of the lamp. The action is cumulative until the lamp is bright enough to lock the SCR into full conduction and remain in this state until the lamp is extinguished. Many alarm devices are using this system.

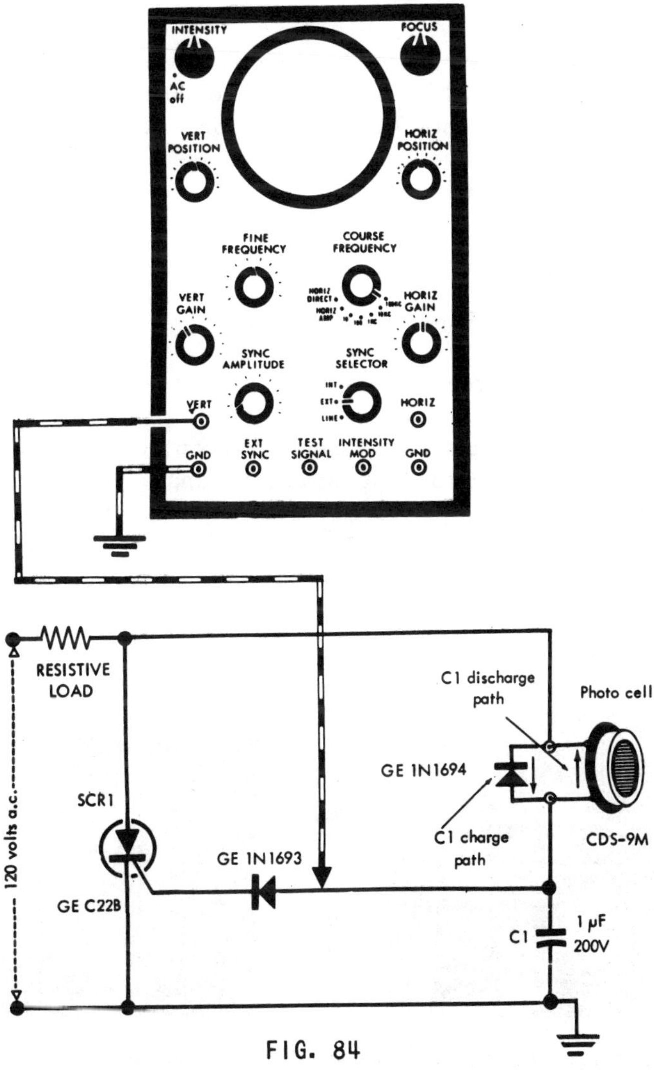

FIG. 84

85

Unijunction Transistor Tests

Unijunction transistors are now being used in many commercial applications and will eventually require service. It is not advisable to test a unijunction transistor with an ohmmeter; therefore, a test method must be devised. Here, the oscilloscope can be of great help.

Procedure

Equipment Required:

Oscilloscope, direct probe

The adapter in Fig. 85 is simply a relaxation oscillator consisting of two resistors, a capacitor, and a 6-volt battery. It is actually a free-running sawtooth generator operating at a frequency determined by the time constant of R1-C1 (R1 is variable). Notice that C1 is connected between the emitter and base 1. When the capacitor charge reaches a critical

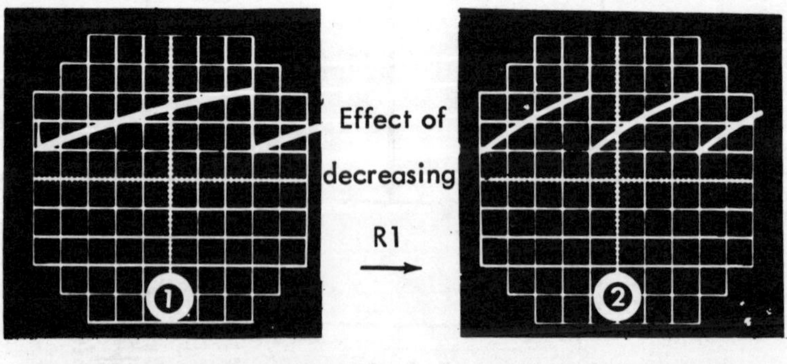

Horizontal sweep calibrated to one millisecond per square

value the unijunction transistor "fires" and the emitter conducts while the capacitor discharges rapidly through the low resistance path between the emitter and base 1. This completes one cycle of charge and discharge. Connect the oscilloscope vertical input leads across C1 and vary R1 while observing sawtooth linearity and response. If the vertical amp is calibrated, it is possible to measure the voltage required to "fire" the unit. See the oscillograms.

FIG. 85

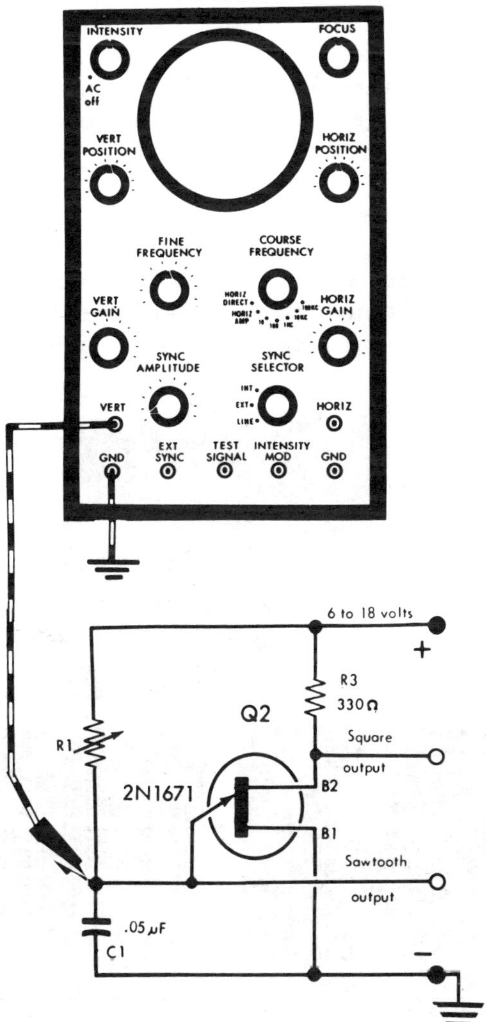

86 Transistorized "RC" Circuit Checks

When the capacitor charge current is held constant in an RC circuit, the integrated voltage rises linearly. Therefore, a linear charge rate is facilitated by using a component with a constant-current characteristic. For example, a junction transistor is an ideal replacement for the fixed resistor in an RC circuit because the collector current remains constant regardless of any variation in collector voltage. As the capacitor is charging, the collector voltage is decreasing; despite this, the collector current remains constant, providing the base bias current remains fixed. See Fig. 86. The internal resistance between the collector and emitter constitutes a series resistance and, as such, becomes a major factor in the RC time constant of the charge path. A variation of the base current will vary this resistance, therefore a base

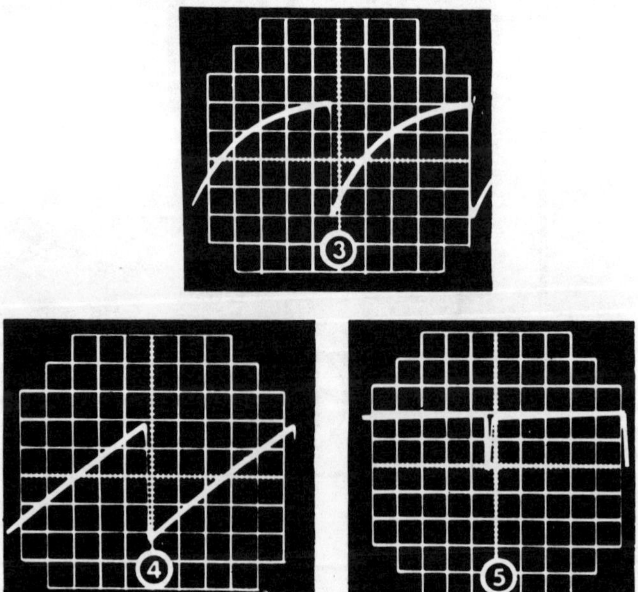

current control will serve as a "fine frequency" adjustment without changing the amplitude of the sawtooth output.

Oscillogram 3 shows the output waveform when transistor Q1 is replaced with a resistor. Oscillogram 4 is the output waveform with Q1 in the circuit. Notice the improved linearity of the sawtooth due to constant current. Oscillogram 5 is the voltage waveform across emitter resistor R2. Oscilloscope vertical sensitivity is 10 millivolts per division. The flat portion represents the constant-current charge period.

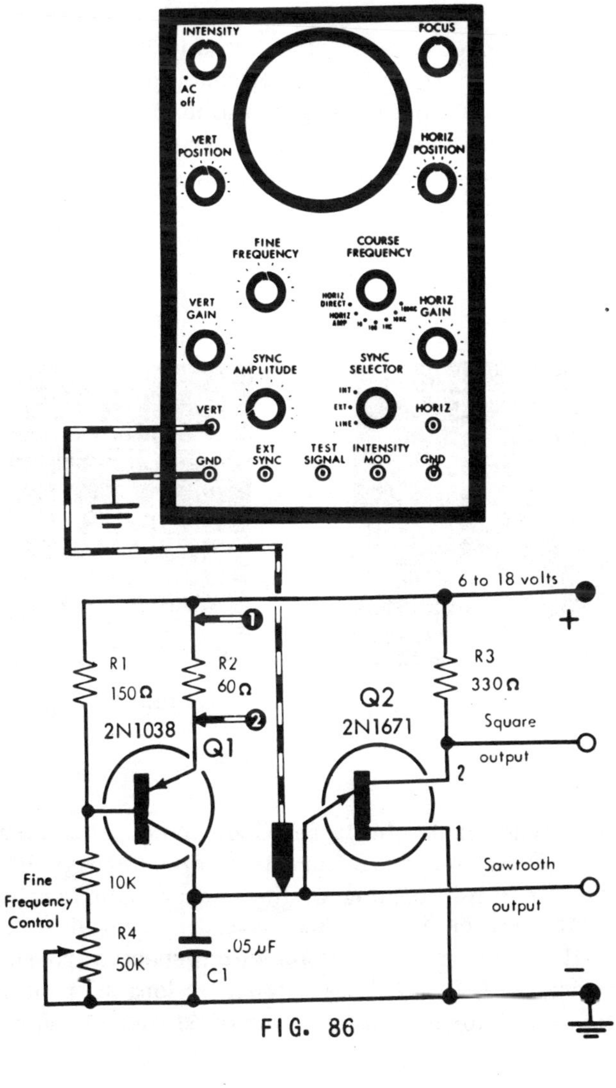

FIG. 86

87 Trigger Generator Waveforms

A unijunction transistor oscillator that differs somewhat from the relaxation oscillator is illustrated in Fig. 87. The difference is in the charge sequence of capacitor C1. Notice that the charge path is from the negative line via base 1, emitter, and R3 to the positive line. When power is applied the voltage distribution across the charge path is such that it develops a forward bias across the PN junction. This causes an emitter current to flow through the charge path until capacitor C1 is

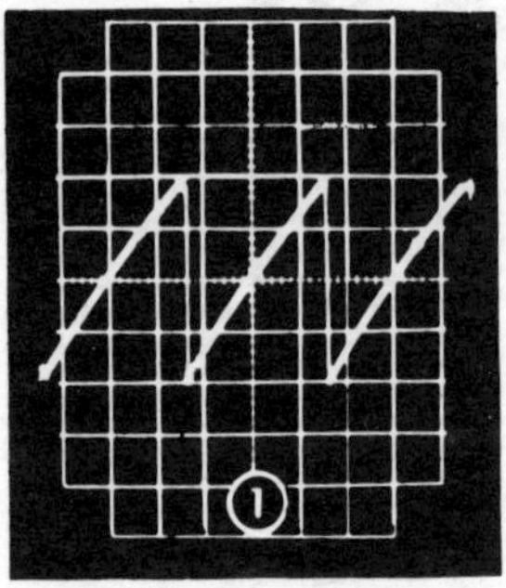

C1 slow discharge

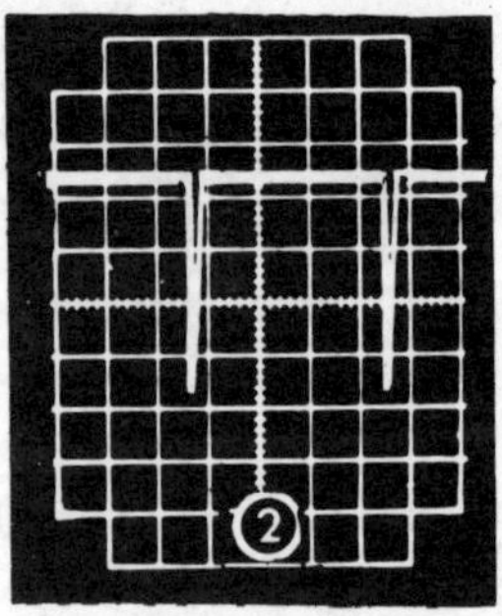

C1 rapid charge

charged. When this point is reached, the emitter current is almost zero and the capacitor discharges through R1, R2, and R3. The time constant of the charge path is relatively short. C1 charges rapidly and develops a negative trigger pulse at the junction of C1 and R3 with respect to ground. The time constant of the discharge circuit is longer and develops a sawtooth waveform at the junction of C1 and R1; during this time the emitter is cutoff. Comparing this oscillator with the relaxation type, it is interesting to note that the transistor is fired first and its emitter current charges C1. This is a reverse procedure to the relaxation oscillator.

FIG. 87

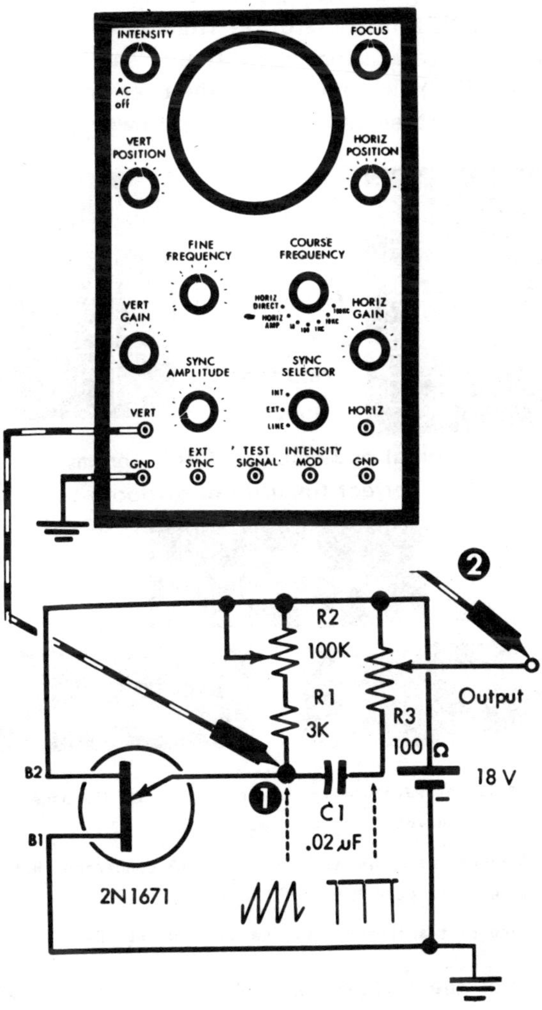

88

Superlinear Sawtooth Generator Waveforms

A linear sawtooth generator circuit using a unijunction transistor and an NPN transistor is illustrated in Fig. 88. Q1 operates as a relaxation oscillator. The sawtooth output (emitter terminal) is connected to the base of an NPN transistor which operates as an emitter-follower buffer stage.

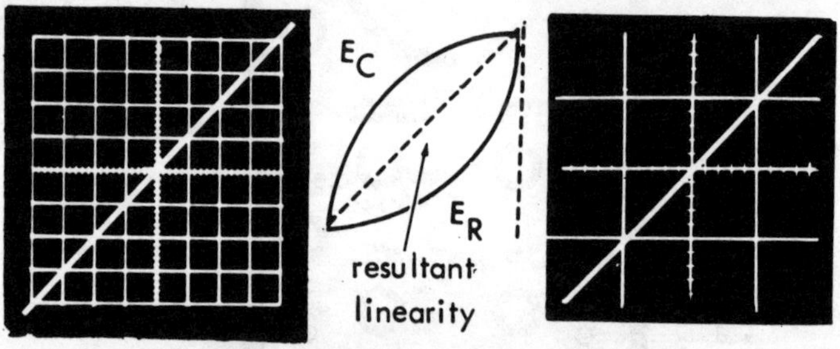

Original and enlarged Oscillograms show perfect linearity of sawtooth.

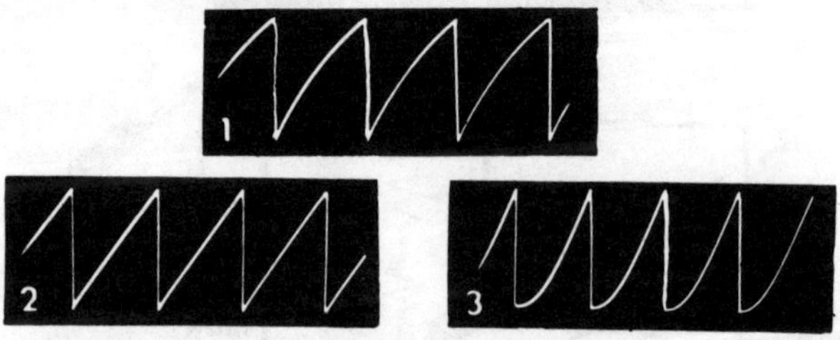

1. Sawtooth waveform at output with C3 disconnected. Notice curvature of this waveform.
2. Sawtooth waveform at output with C3 connected. Notice absence of curvature.
3. Incorrect adjustment of feedback control R4.

A sample of the integrator voltage is taken off at C2 and applied through R4 to load resistor R6. This provides an integrator type feedback to compensate for the loading of the output stage. Feedback control R4 is adjusted for optimum linearity, augumented by bootstrap circuit R3 and C3 which improves linearity. The waveforms shown in Oscillograms 1 and 2 illustrate a sawtooth output with the bootstrap circuit disconnected (1) and with it connected (2). Notice the decided boost in amplitude and elimination of the curvature. The bootstrap potential applied to the emitter is a differentiation voltage and is falling, while the potential applied to the base of Q2 is an integrator voltage and rising. The two combine at the emitter of Q2 (see the diagram).

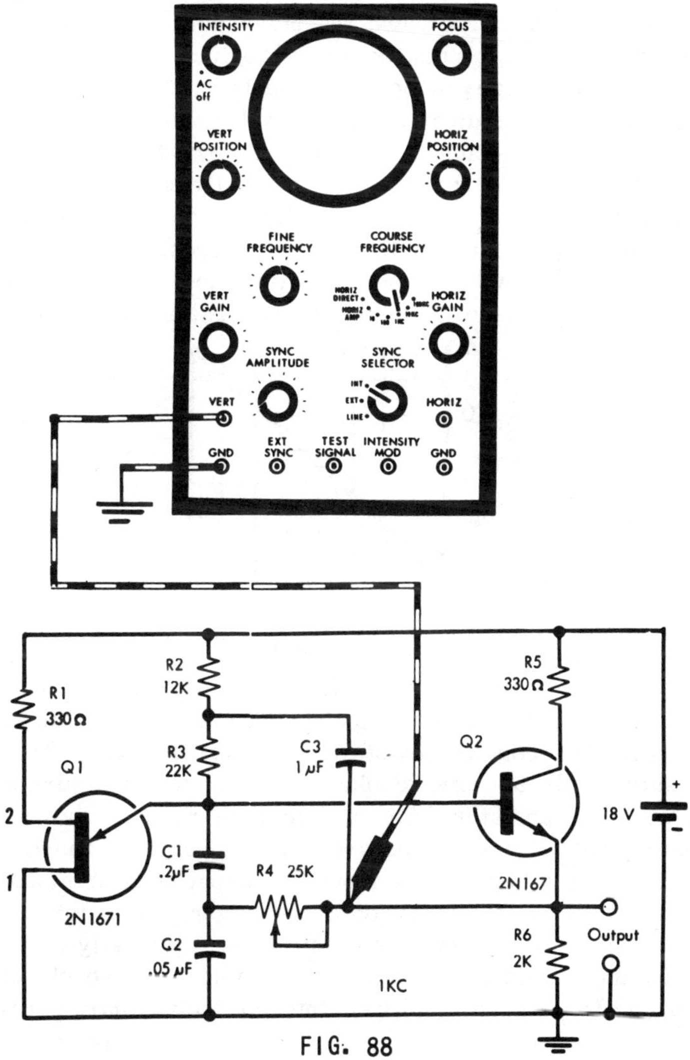

FIG. 88

89 Pulse Delay Control Measurements

A pulse-delay control circuit, using a monostable unijunction transistor, is illustrated in Fig. 89. The time delay is a function of capacitor C1 and the delay period is dependent upon its capacity. When power is applied, the voltage to which C1 can charge is dependent upon the voltage ratio across R1 and R2. The circuit is designed so that the voltage across the R1/C1 combination is sufficient to fire the unijunction transistor. Hence, the circuit is normally cut off.

A positive pulse of sufficient amplitude applied to the input charges C1. This boosts the voltage across the capacitor sufficiently to forward-bias the emitter and fire the transistor. When this occurs, D1 is reversed biased (cut off)

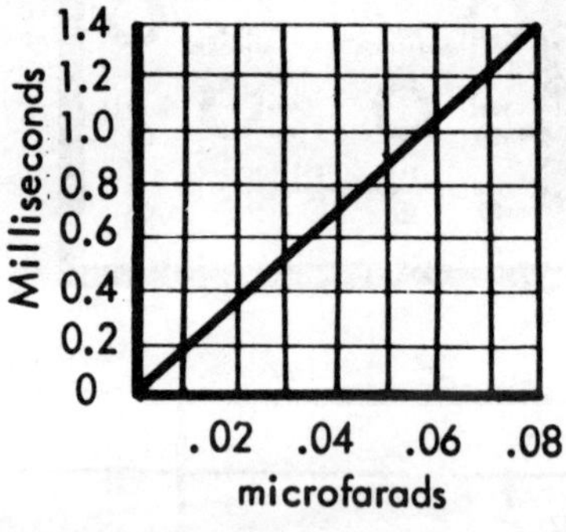

and an electron current flows from ground via base 1, the emitter junction, and R2 to the positive terminal of the supply line. This condition exists until C1 has discharged through R1 sufficiently to forward-bias diode D1, which in turn will cut off the transistor. The circuit then remains quiescent (resting) until another positive pulse appears at the input. The time delay between the input and output is determined by C1. The relationship is given in the accompanying table.

A pulse delay system has many uses, especially in color television circuitry where it is used to delay a flyback pulse in order to gate an amplifier into conduction during trace time.

FIG. 89

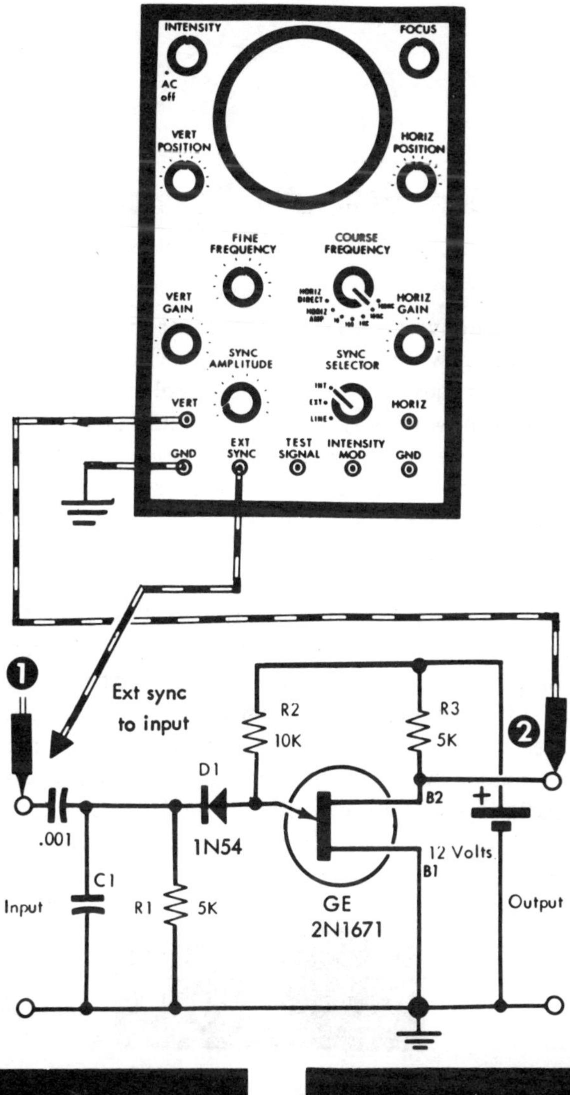

Original pulse

Delayed pulse

90 Step Generator Waveforms

A circuit that converts a trigger generator output to a staircase wave is shown in Fig. 90. This circuit has many applications such as frequency division, control functions, etc. The negative trigger pulses applied to the base of PNP transistor Q2 drive it into periodic conduction. Capacitor C2, connected in series with the collector, stores up each pulse of collector current which appears as a distinct step of voltage at the collector terminal with respect to ground. As each pulse arrives it adds an additional step to the collector voltage, and this buildup will continue until a "staircase" pattern appears at the collector of Q2. When the voltage across C2 reaches the firing point of Q3, the unijunction transistor fires and C2 discharges, after which a repeat set of stairs takes place.

R4 varies the number of steps in each staircase cycle. For example, decreasing R4 reduces the amplitude of the trigger input to Q2, therefore more steps may be accommodated per staircase. R2 varies the frequency of the staircase cycle.

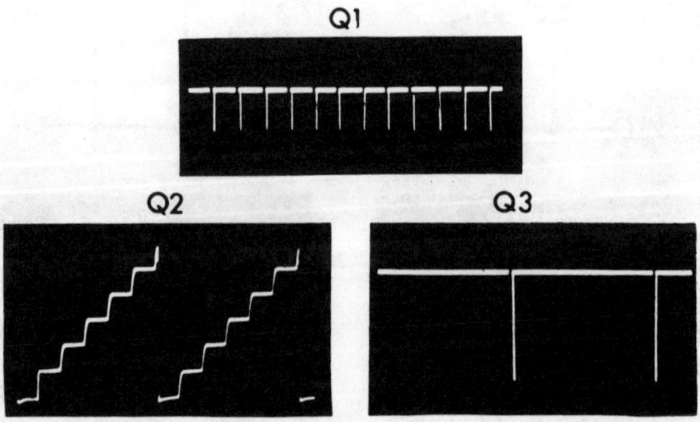

Twelve pulses from Q1 build two staircase cycles at collector of Q2 providing two pulses at output of Q3.

For example, increasing R2 extends the discharge period of C1 and reduces the number of trigger pulses. With fewer trigger pulses per second, it will take longer to build a staircase pattern. The generator may be used as a frequency divider or used to trigger another circuit at any predetermined step level. Frequency division is a feature that may be extended to additional stages similar to Q2 and Q3. It has been used in conjunction with a transistor curve tracer to produce steps of base current, thereby producing a family of curves for simultaneous observation.

FIG. 90

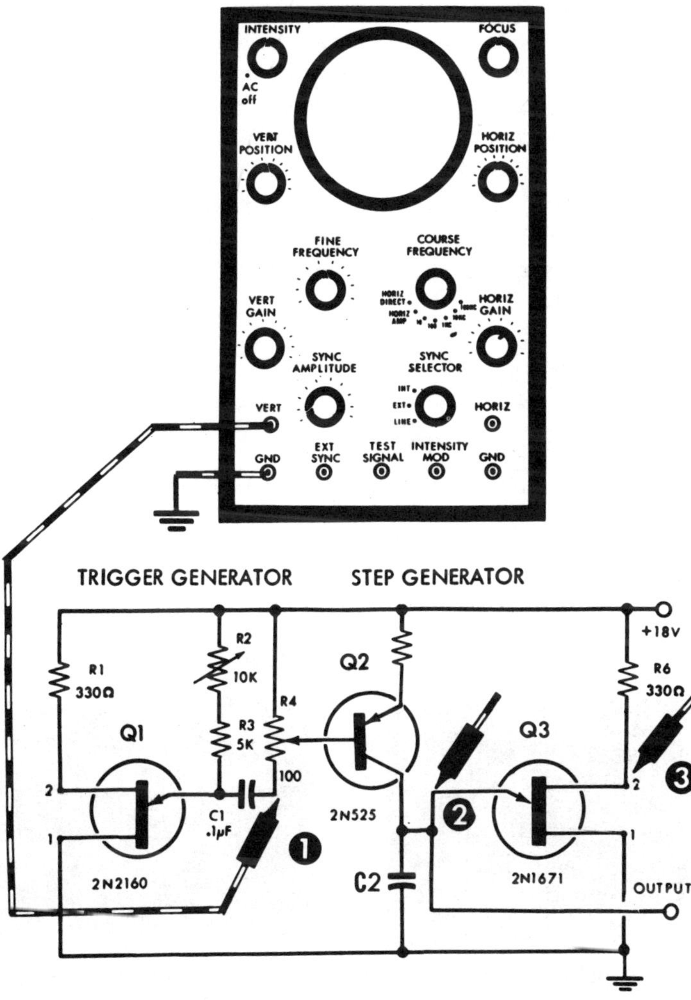

91 Tunnel Diode Oscillator Checks

Vacuum tube and transistor oscillators use a third element (the control grid of a vacuum tube or the base of a transistor), but a tunnel diode has no third element as such, rather a "quantum mechanical" tunneling of energy through the diode junction. For example, while the tuned tank circuit is oscillating, its alternating voltage appears across the diode, aiding and opposing the DC bias across it. Hence, the diode bias increases and decreases in unison with the frequency of the tuned circuit. When the diode bias decreases, it supplies additional current in accordance with its negative resistance characteristic; i. e., the lower the bias voltage the higher current.

When a bias of .125 volt is applied to the diode, a tunnel current of about .5 ma will flow. To apply this small voltage

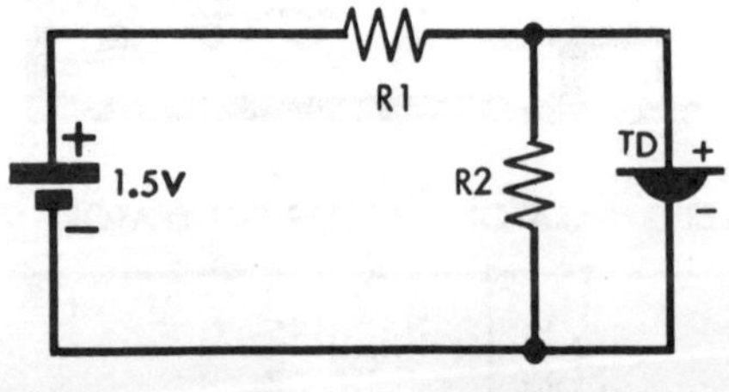

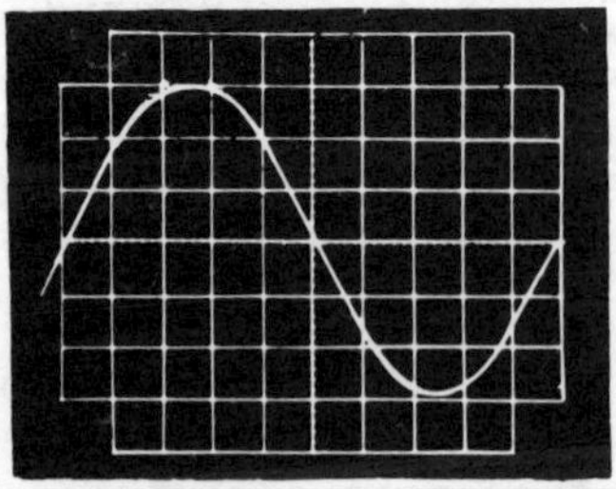

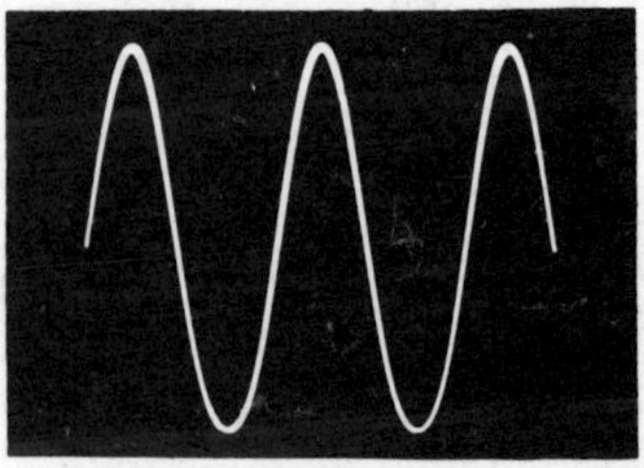

Undistorted sinewave from tunnel diode oscillator

from a 1.5-volt battery requires a series resistance of 3000 ohms. However, the negative resistance of the diode is 100 ohms; therefore, the total series resistance is 2900 ohms. This would obviously cancel the negative resistance effect of the tunnel diode. To avoid this, the bias voltage should be developed across a resistance value that is less than the negative resistance of the diode. For example, by using a 20-ohm bias resistance and connecting the diode in parallel, the combined resistance will be 80 ohms. See Fig. 91.

FIG. 91

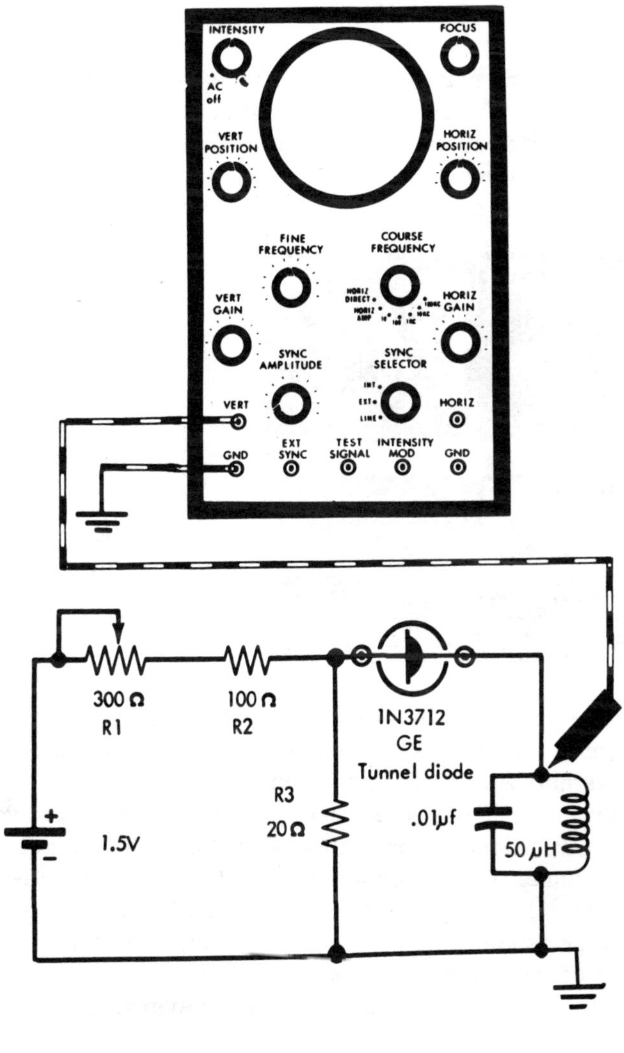

92 Full-Wave Phase Control Checks

A bi-directional silicon controlled rectifier operating with a single gate electrode is illustrated in Fig. 92. This circuit uses a GE unit called a triac. Connected in the gate circuit is a bi-directional double diode unit, called a diac, which is used to equalize the timing of the gate trigger pulses during the positive and negative halves of the applied cycle. When power is applied C1 starts to charge, and when its voltage reaches a critical value the diac conducts and triggers the triac. At this moment C1 partially discharges through the diac and the gate circuit of the triac. After this initial trigger

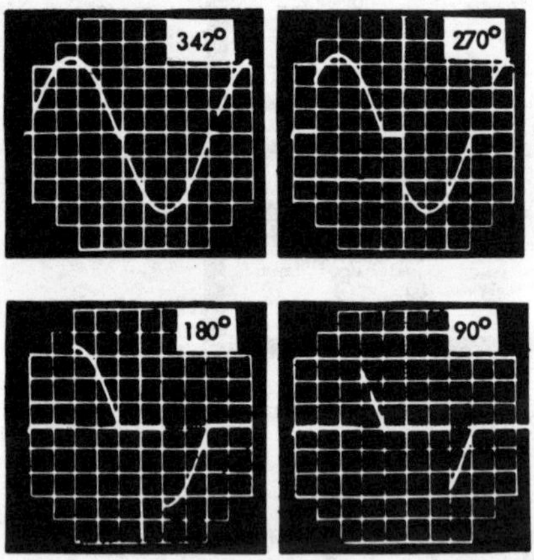

CONDUCTION ANGLE FOR FULL WAVE SCR

OSCILLOSCOPE CONNECTED ACROSS THE LOAD

action the triac continues to conduct for the remainder of the half cycle. The trigger action occurs in both the positive and negative half waves of the applied cycle.

Varying R1 shifts the phase of the sine-wave voltage across C1, which in turn shifts the phase of the trigger pulse. In this manner the trigger action may be delayed or advanced, thus controlling the power applied to the load. Notice the oscilloscope is connected across the load, hence the working portion of the cycle is displayed vertically. With the oscilloscope connected across the triac the reverse is true. The oscillograms show the full wave conduction angle, calibrated at 45° per division. If each half cycle shows a conduction angle of 90° at full wave the total conduction angle would be 180° (see the oscillogram representing 180°).

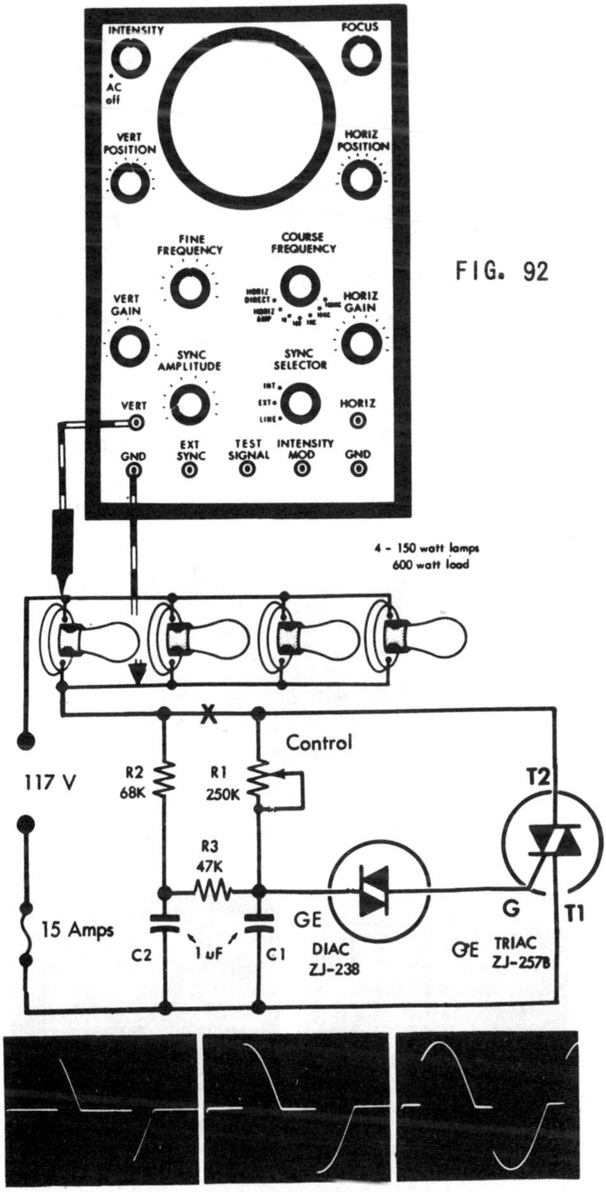

FIG. 92

FULL WAVE PHASE CONTROL

93

Modified Halfwave SCR Control Waveforms

An SCR circuit with smooth continuous control for loads up to 600 watts is illustrated in Fig. 93. It operates with maximum efficiency with practically no losses and may be used to control lighting, universal motor speeds, heating, and DC power supplies. During the negative half wave of the input cycle, the SCR is off and C1 charges rapidly via diode D1 and discharges via R1 and R2. However, the rate of discharge is governed by the adjustment of R2 which functions as a power output control.

When R2 is adjusted to zero ohms, the time constant of the discharge path is short and C1 charges and discharges with

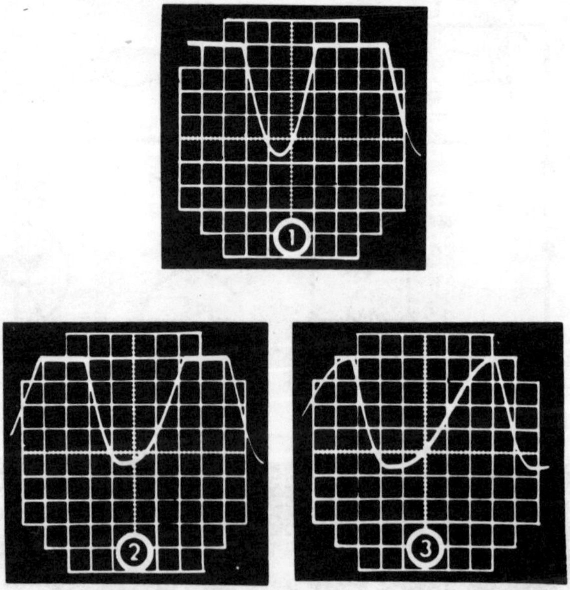

Capacitor charge and discharge

the negative swing of the input cycle. When this condition prevails C1 is discharged and the gate potential is zero at the beginning of the positive half wave. This causes the SCR to conduct during the positive swing of the input cycle, (conduction angle 180°) which delivers maximum power to the load (see Oscillogram 1). Gradually increasing R2 increases the time constant, delays the discharge of C1, and the gate potential is held negative for a longer period. When this condition prevails a portion of the positive half wave is used to overcome this effect and the SCR conducts for less than 180°,

FIG. 93

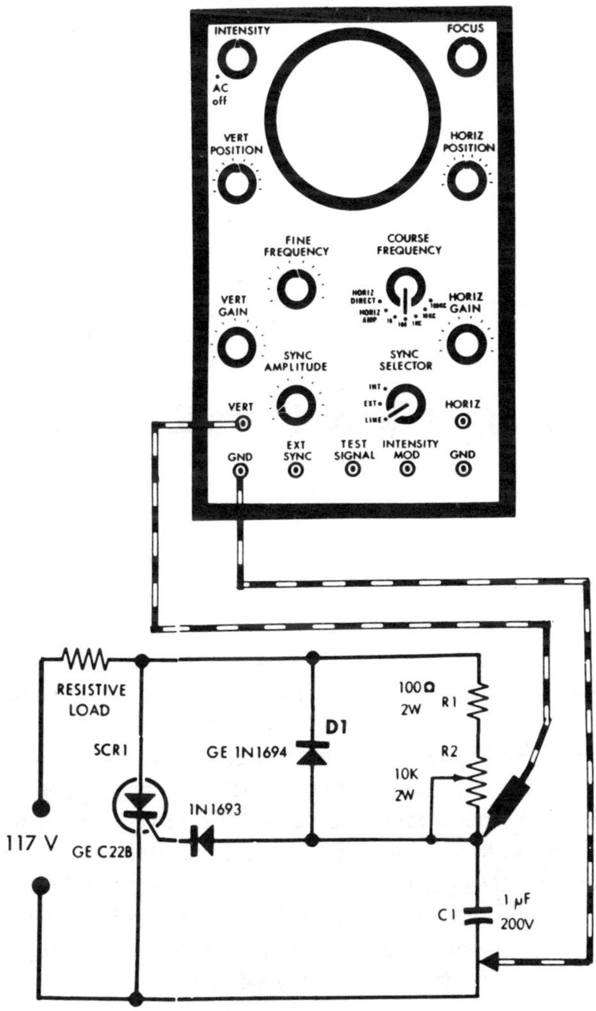

thereby reducing the power delivered to the load. Oscillogram 2 shows a delay of one division or 60°. In Oscillogram 3 the delay is three divisions or 180°; i.e., no conduction. When R2 is adjusted to maximum (10K) the discharge time constant is very long and the SCR remains off. Therefore, when R2 is adjusted from maximum to minimum it provides a smooth control of power output from zero to maximum, thus representing a conduction angle variation from 0 to 180°. Notice in Oscillogram 1 that the half cycle engages three divisions and is delayed for an additional three divisions in Oscillogram 3. In Oscillogram 3 the positive half wave has been completely shut off. In Oscillograms 1 and 2 the flat portion of the trace indicates SCR conduction.

---

94 Triangular Wave Generator Tests

Square waves and sine waves are employed in various applications, separately or combined, in modern electronics. However, the triangular waveform (back-to-back sawtooth) has come into prominence and is often used in sweep operation, timing circuits, and numerous other uses. Fig. 94 illustrates the circuit used to generate a triangular waveform of exceptional linearity with a constant amplitude, regardless of frequency. The linear rise and fall of the waveform provides an excellent means for observing distortion and is far more effective than the sine-wave test, since any irregularity is easily detected. The adjustment of R in the emitter circuit determines the frequency, while the amplitude is held constant by the precise firing time of the transistor.

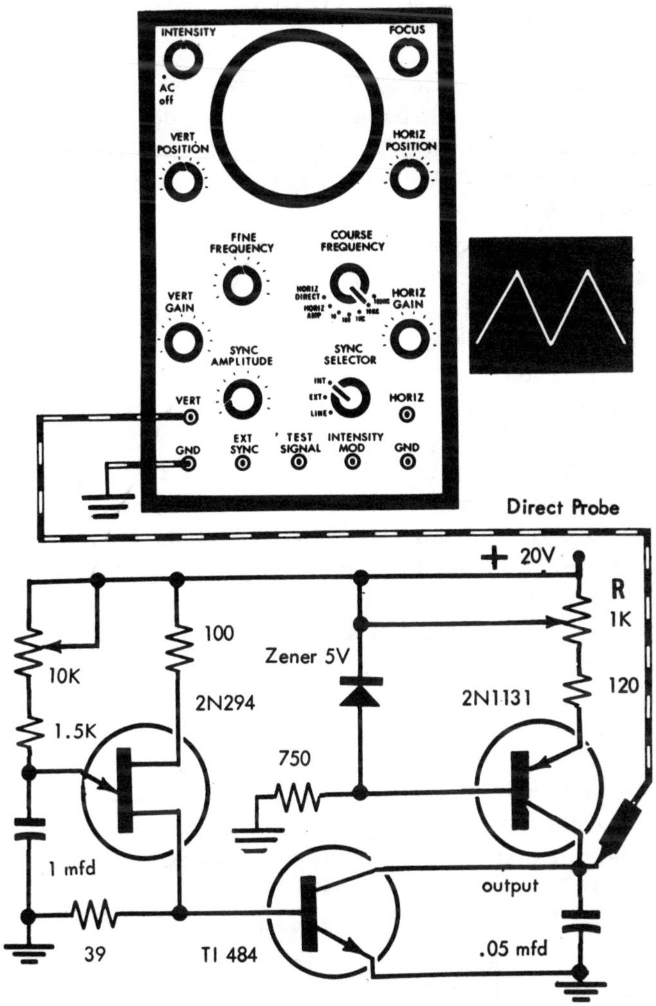
INTENSITY
FOCUS
AC
off
VERT
POSITION
HORIZ
POSITION
FINE
FREQUENCY
COURSE
FREQUENCY
VERT
GAIN
HORIZ
GAIN
SYNC
AMPLITUDE
SYNC
SELECTOR
INT
EXT
LINE
VERT
HORIZ
GND
EXT
SYNC
TEST
SIGNAL
INTENSITY
MOD
GND
Direct Probe
+ 20V
R
1K
100
Zener 5V
10K
120
2N294
2N1131
1.5K
750
1 mfd
output
39
TI 484
.05 mfd

FIG. 94

95

Tunnel Diode Multivibrator Waveforms

When power is applied to the multivibrator circuit in Fig. 95 C1 begins to charge through R2 and R3, developing a bias voltage across the tunnel diode. (C1 is connected across R1 and the tunnel diode.) When the capacitor voltage rises above zero, a current will flow through the tunnel diode and R1. When the diode current reaches peak value, the diode switches from a high-current, low-voltage state (low impedance) to a low-current, high-voltage state. This occurs in the negative resistance region of the curve. See diagrams A and B. The switch from a low to a high impedance diverts the current flow through the emitter-base junction of the transistor, and in turn causes this stage to conduct.

Simultaneously, the capacitor discharges through the conducting transistor and R2. During the discharge of the capacitor, the voltage across the tunnel diode decreases and the diode returns to its low-impedance state. This, in turn, diverts the emitter-base current allowing the transistor to go into cutoff. The capacitor starts to recharge and the cycle of events is repeated. Notice that it is the charge and discharge of C1 that causes the tunnel diode to gate the transistor on and off. The repetition rate of the square wave is dependent upon the RC time constants of the charge and discharge paths of the capacitor. Actually when the tunnel diode is in its low-impedance state it short circuits the base-emitter junction of the transistor and causes this stage to cut off.

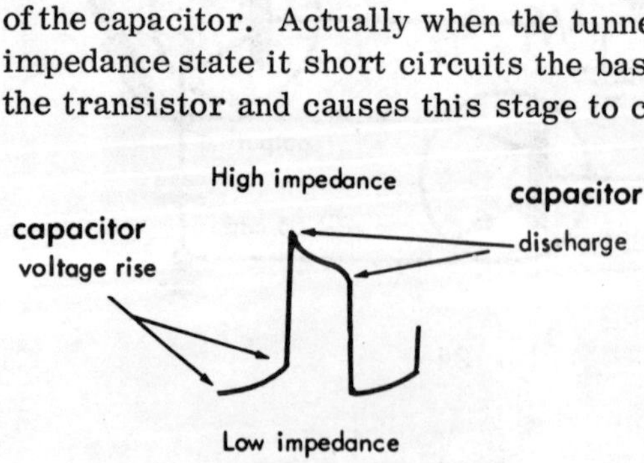

Voltage waveform across diode

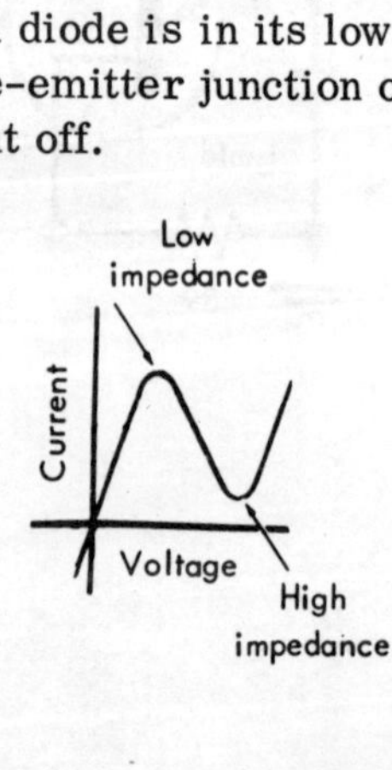

FIG. 95

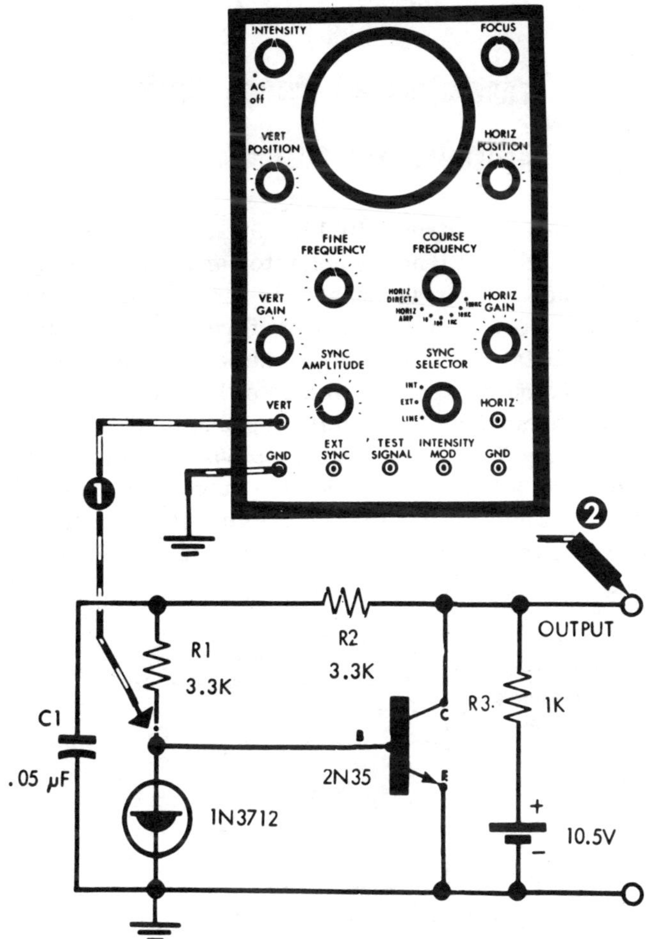

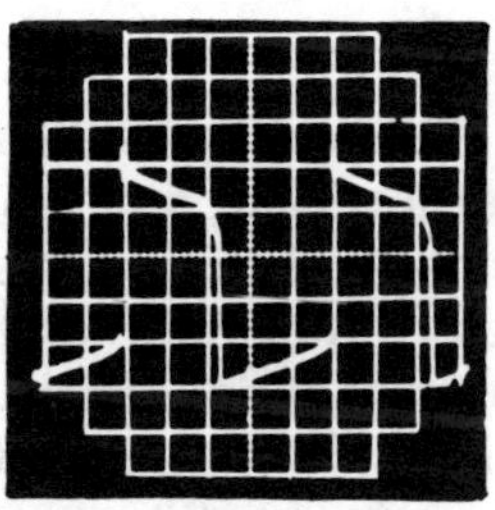

Voltage waveform across diode

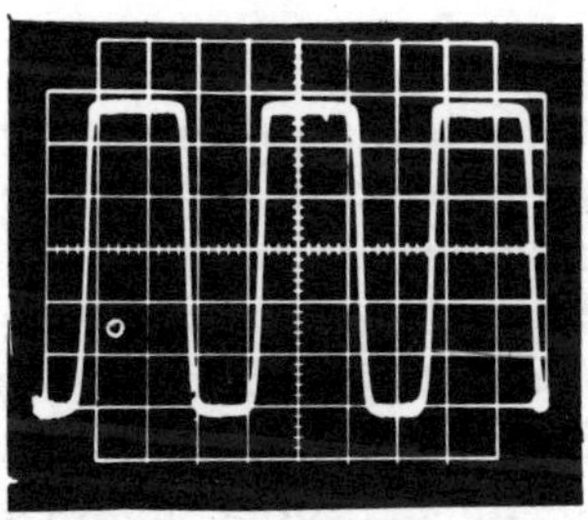

OUTPUT

96

Transistorized Phase Shift Circuit Measurements

The one-transistor circuit in Fig. 96 functions as a phase-shift circuit. Notice that the collector and emitter resistors are equal, hence the stage has near unity gain, making the circuit equivalent to a 1-to-1 transformer. The amount of phase shift obtained at the output network is determined by the applied frequency and the values of R5 and C2. The circuit shown is capable of producing a phase shift from zero to 180° by varying R5. For example, a 50K potentiometer will vary the phase a full 180°. Since the stage has a paraphase amplifier characteristic, the input signal is inverted 180° at the collector terminal and no phase shift is introduced at the emitter.

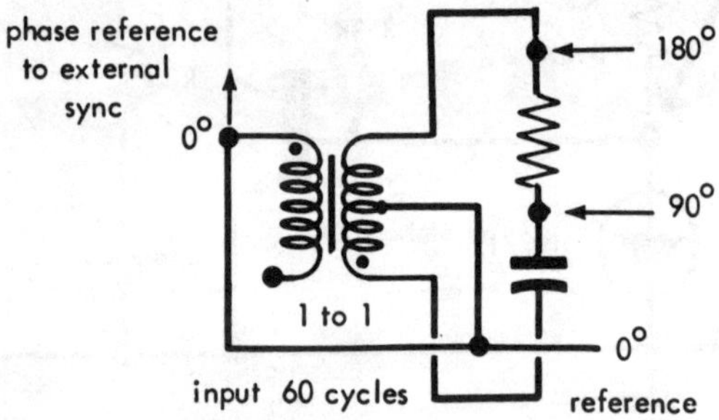

Procedure

Set up the test as shown in Fig. 96. Connect the oscilloscope external sync terminal to the amplifier input terminal and switch the sync selector to the "ext sync" position. (This lead should remain connected during the test.) Adjust the horizontal sweep so that a single cycle engages 8 divisions, which is 45° per division (see oscillograms). Use a 50K potentiometer in place of R5 and notice the waveform shift when this control is varied. The phase shift comparision may be observed by connecting the vertical input probe to the input and then to the output.

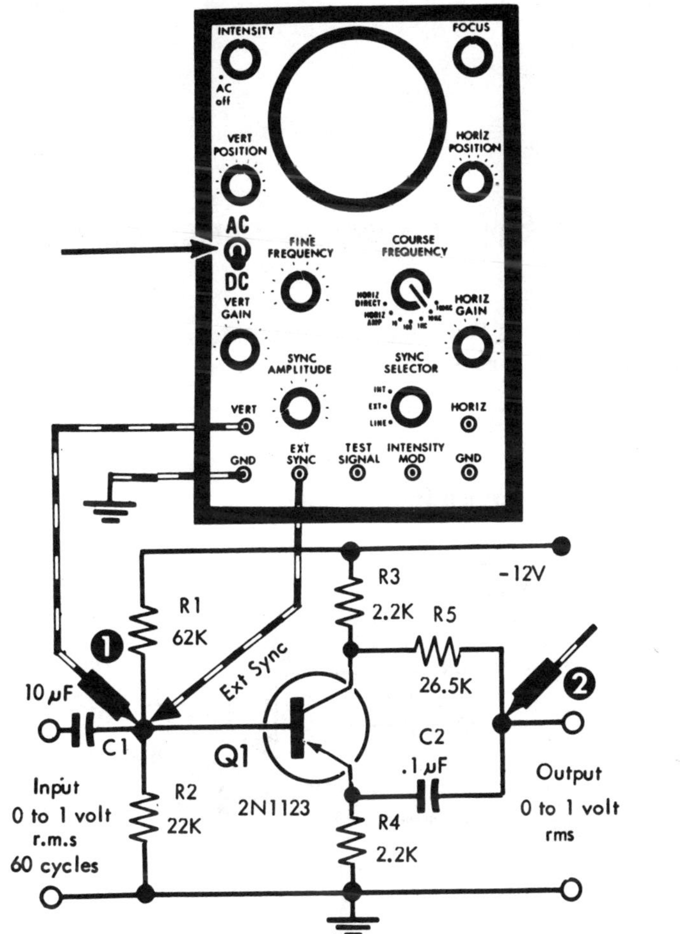

FIG. 96

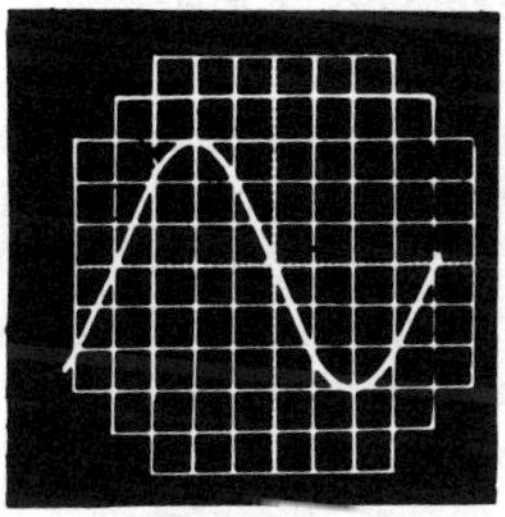

Input phase

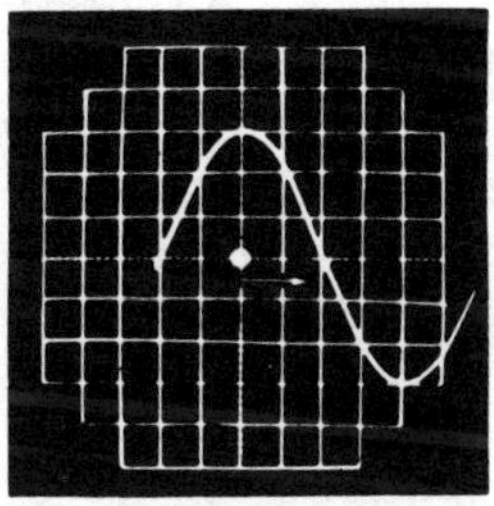

Output phase lagging 90°

Calibration 45° per square

97

Sine-to-Square Wave Converter Waveforms

The circuit in Fig. 97, equivalent to a two-stage Schmitt trigger, efficiently converts sine waves to square waves.

Procedure

Equipment Required:

Oscilloscope, AC and DC amplifiers
Audio signal generator
Carbon resistor—10K

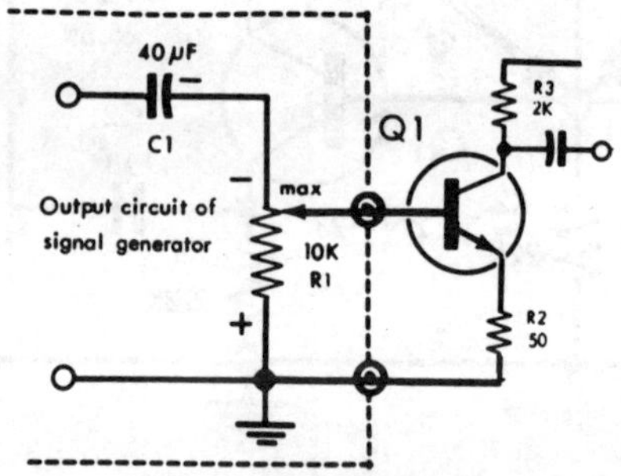

Connect the test as shown in Fig. 97 and adjust the oscilloscope to display a 3-cycle pattern (see Oscillogram 1). Connect the generator output to the transistor input and the transistor output to the vertical input. Observe the clipped positive peaks in Oscillogram 2. During negative peaks the NPN transistor is cut off and provides the flat top of the output wave. See Oscillogram 3. The 10K resistor connected in series with the base lead is used to prevent clamping action due to the diode effect of the base-emitter junction and the output circuit of the signal generator. To observe this effect connect the vertical input of the oscilloscope to the base terminal and short-circuit the 10K resistor (see Oscillogram 4).

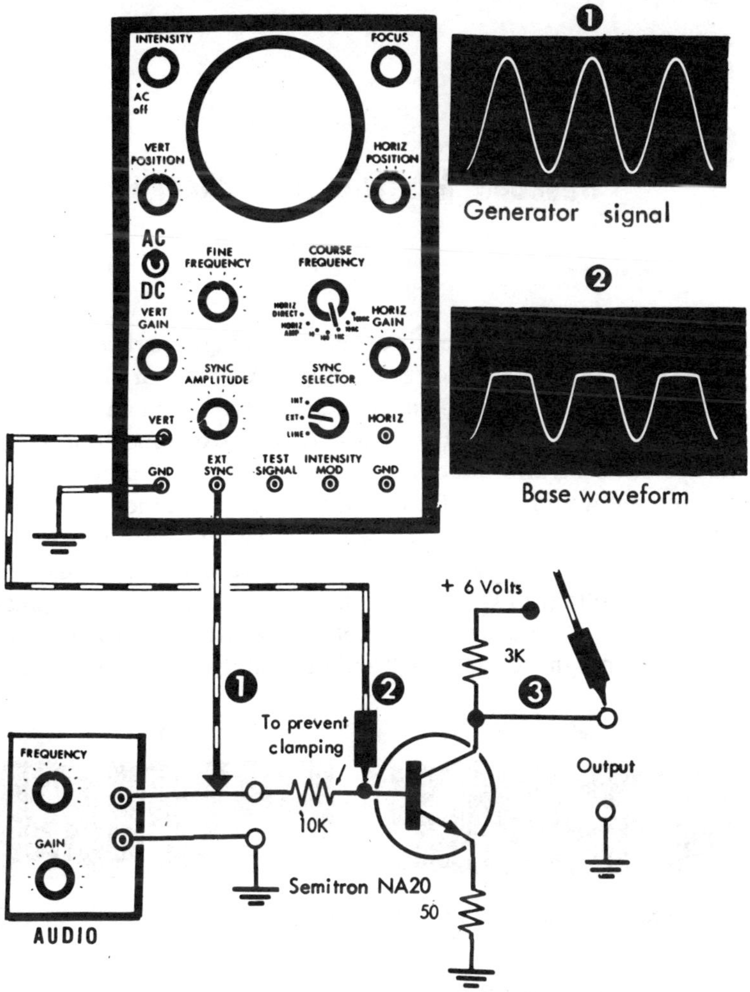
INTENSITY
FOCUS
AC
off
VERT
POSITION
HORIZ
POSITION
AC
DC
FINE
FREQUENCY
COURSE
FREQUENCY
VERT
GAIN
HORIZ
GAIN
SYNC
AMPLITUDE
SYNC
SELECTOR
VERT
HORIZ
GND
EXT
SYNC
TEST
SIGNAL
INTENSITY
MOD
GND
1
Generator signal
2
Base waveform
+ 6 Volts
3K
1
2
3
To prevent
clamping
Output
FREQUENCY
GAIN
AUDIO
10K
Semitron NA20
50

FIG. 97

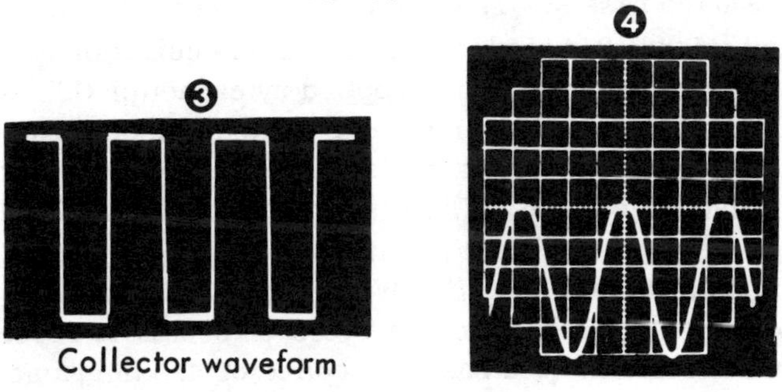
3
Collector waveform
4

98 Cascode Multivibrator Checks

In a cascoded type multivibrator, two stages are connected in series across the power supply. A study of the circuit shown in Fig. 98 would seem to indicate that such an arrangement would rule out its use as a multivibrator since one stage has to cut off while the other is conducting, and a series arrangement would make this impossible. A further study reveals how it oscillates as a free-running multivibrator.

When power is applied, C1 and C2 charge rapidly, and the collector voltage of Q2 charges C3 via the base-emitter junction of Q2. This charge current constitutes a forward-bias current through the base of Q2 and causes the stage to conduct.

❶

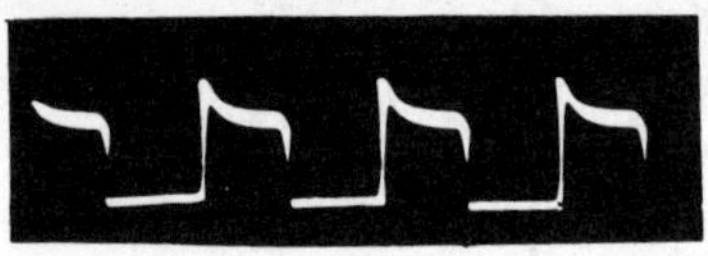

Waveform at the base of Q2

❷

Waveform at the collector of Q2

This, in turn, causes the collector voltage of Q2 to drop and C4 to discharge through R3 and R4. The drop across R4 is positive at the base of Q1, causing this stage to cut off. Now, Q1 is cut off and Q2 is conducting. When this condition occurs, the collector current of Q2 is supplied by capacitor C2. When C4 has discharged sufficiently, Q1 comes out of cutoff and conducts. This causes its collector voltage to drop and, in turn, discharges C3. The discharge current of C3 flows through R1 and develops a positive voltage at the base of Q2, causing this stage to cut off. When C3 has discharged sufficiently, Q2 conducts and the cycle of events is repeated. (The waveforms at Q1's base and collector are the same.)

FIG. 98

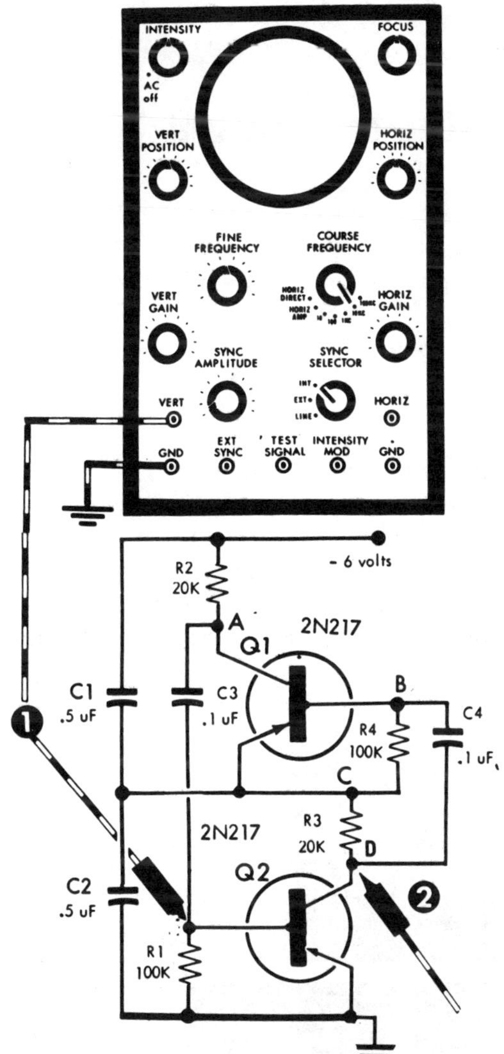

INTENSITY
AC
off
FOCUS
VERT
POSITION
HORIZ
POSITION
FINE
FREQUENCY
COURSE
FREQUENCY
VERT
GAIN
HORIZ
GAIN
SYNC
AMPLITUDE
SYNC
SELECTOR
INT
EXT
LINE
VERT
HORIZ
GND
EXT
SYNC
TEST
SIGNAL
INTENSITY
MOD
GND
- 6 volts
R2
20K
A
Q1
2N217
B
C1
.5 uF
C3
.1 uF
C4
.1 uF
R4
100K
C
R3
20K
D
2N217
Q2
C2
.5 uF
R1
100K
1
2

99

Series-Connected Schmitt Trigger Tests

The conventional Schmitt trigger circuit always has one stage conducting despite the on or off state of the circuit. The new Schmitt trigger circuit shown in Fig. 99 uses fewer parts. And when the circuit idles, both stages are cut off.

Procedure

Transistor Q1 is an NPN and Q2 is a PNP. When the input sine wave from the generator is zero, Q1 and Q2 are cut off. This condition exists until the input voltage rises above zero and forward biases the base of Q1. When this occurs, the collector current of Q1 develops a voltage drop across R2, which forward biases Q2 and drives this stage into conduction.

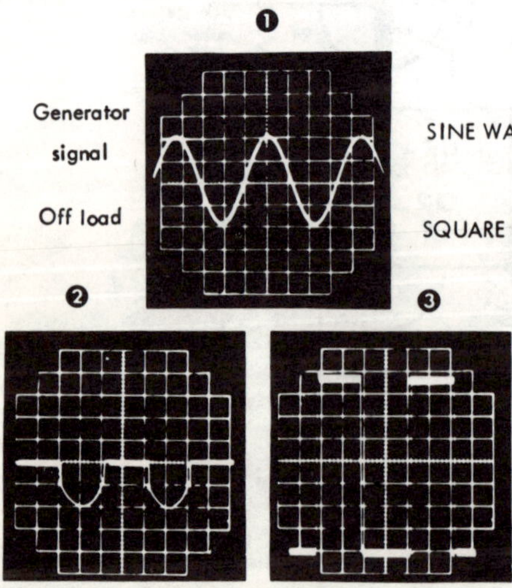

Base waveform

Collector waveform

1. Signal generator output
2. Signal at the input of Q1. Notice clipped positive peaks due to conduction of Q1
3. Signal at the output of Q1. Notice steep leading and trailing edges

As the input voltage to Q1 increases, both transistors saturate. Notice the collector output of Q2 is connected to the input of Q1. This unique arrangement causes both transistors to saturate and cut off simultanously to produce a square wave with steep leading and trailing edges at a repetition rate corresponding to Q1's input frequency. During the negative swing of the input cycle, the whole circuit remains in an off condition. When Q2 is removed, the positive half cycle reverts to its original shape.

FIG. 99

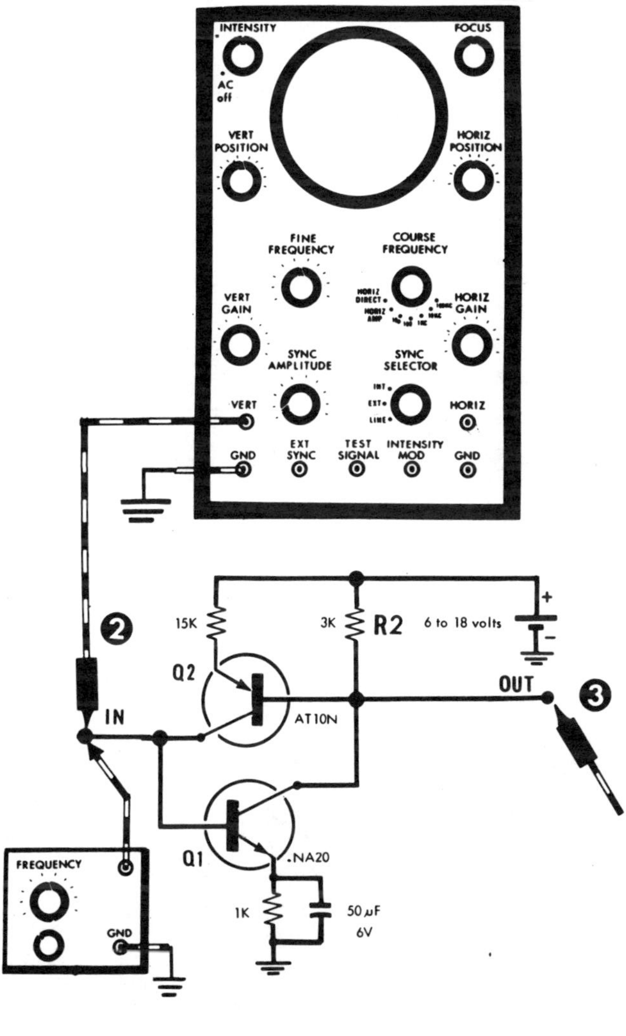